Bath and West of England Society

Letters and Papers on Agriculture and Planting, etc

Vol. IV.

Bath and West of England Society

Letters and Papers on Agriculture and Planting, etc
Vol. IV.

ISBN/EAN: 9783337125226

Printed in Europe, USA, Canada, Australia, Japan

Cover: Foto ©berggeist007 / pixelio.de

More available books at **www.hansebooks.com**

LETTERS AND PAPERS

ON

Agriculture, Planting, &c.

SELECTED FROM

THE CORRESPONDENCE

OF THE

Bath and West of England Society

FOR THE ENCOURAGEMENT OF

AGRICULTURE,	MANUFACTURES,
ARTS,	AND COMMERCE.

VOL. IV.
THE SECOND EDITION.

BATH, PRINTED, BY ORDER OF THE SOCIETY,
BY R. CRUTTWELL;
AND SOLD BY C. DILLY, POULTRY, LONDON,
AND BY THE BOOKSELLERS OF BATH, BRISTOL, SALISBURY,
GLOCESTER, EXETER, &c. &c.

M DCC XCII.

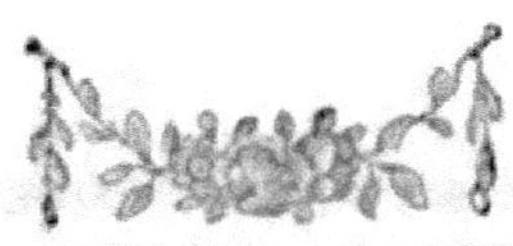

C O N T E N T S

OF

VOLUME IV.

DIRECTIONS FOR PLACING THE PLATES.

INTRODUCTION.

INTRODUCTION.

THIS Society having, in its Book of Premiums for 1787, announced an intention of ſoon publiſhing a Fourth Volume of Papers; the public had certainly a right to expect an earlier fulfilment of that intention.

For the delay, ſeveral reaſons might be aſſigned; among which, and not the leaſt operative, was the deceaſe of Mr. EDMUND RACK, the late uſeful and ingenious Secretary of this Society. By that event was diſſolved an ideal arrangement of materials; and even for ſome time was ſuſpended, a certainty to whom ſome few of the MSS. had been lent for peruſal.

On adverting, afterwards, to the promiſe of publication, ſome of the moſt active Members of the Society, though agreeing on the propriety of printing a fresh volume, were perhaps less urgent than formerly about the expedition of the measure. The absence of many of the country gentlemen during the last summer, and the greater propriety of determining on the time of printing, when the Committees should be more united, was another reason of a temporary delay. But this interval was not unoccupied with business of a similar nature, and of no inconsiderable importance; which was the superintendance of the press, through the printing a second edition of the former three volumes—all of which are now completely reprinted.

Among the many proofs of the usefulness of this Society, and the public respect which has been paid to its past progress, the complete disposal of the first edition of those volumes, and the increase of the public demand, make at once a convincing and a pleasing testimony. But while the necessity for a new edition of former volumes was flattering, due care to preserve a reputation for usefulness became a new incentive to deliberation and the proper choice of materials. Of this deliberation and care, it is hoped the present volume will furnish some evidence.

The caution which has been used, has given time for remarking the different opinions that have been imparted, by particular gentlemen on the general subject of publication; and it may not be improper here to acquaint the public, that though this Society can never wish to *withhold* those articles of practical communication which it is its plan to *encourage*, yet the importance of *publishing well* appears too great to be hazarded by any arbitrary adherence to stated periods.

On a work like this, written by a variety of hands, and on an almost equal variety of subjects, every man, who in connexion therewith considers the fallibility of human skill, and the varied complexion of the human mind, will anticipate *criticiſm*. Neither from works of this ſort, on which may poſſibly depend the ſucceſs or diſappointment, profit or loſs, of ſubſequent experiment, ſhould the patrons of improvement wiſh to have the ſtrictures of experience and intelligence averted.

An ingenuous and candid reader will not be leſs thankful for a new hint, or a new reſult of experiment, which he finds to be fraught with his own and the public advantage, becauſe it is not unaccompanied in the ſame volume by inequality and imperfection! ——And in return for the liberality of reception which this Society is confident of finding with the public, we think it incumbent upon us to give the following aſſurance, viz. That the ſterling worth of thoſe animadverſions which proceed from real knowledge, and from that zeal for the advancement of truth which this Society has mainly in view, will be ever gratefully diſtinguiſhed from the caſual cavils of inexperience, or irrational attachment to cuſtom.

The

The firſt Article in this Volume, the communication of a Gentleman of large experience in rural ſcience, is long and elaborate. But as the ſubject is confeſſedly of great importance in the preſent improved ſyſtem of cultivation, it is preſumed that no apology can be neceſſary for inſerting, at large, a piece which is evidently the reſult of uncommon ingenuity, and practical obſervation. The value of *Potatoes*, both as an article of huſbandry, and general conſumption in families, is now ſo fully eſtabliſhed, that no argument is required to enforce it. The ſeries of facts, thus given by Dr. ANDERSON, and interſperſed with numerous remarks reſulting from thoſe facts, it is preſumed, will throw very conſiderable light on the properties and value of the root in queſtion. And if ſome experiments ſhould be found of comparatively leſs importance than others, and ſome conjectures of a leſs certain tendency, the Society cannot but be aſſured, that a large tribute of public praiſe will be paid to the

the author of the eſſay:—a man, who, while eminently converſant in the abſtruſe and elegant departments of ſcience, could devote ſo much time and pains to the culture of a ſingle root; but a root on which the ſubſiſtence and comfort of a large part of the poor of theſe kingdoms has been known to depend.

The piece on the ſubject of a *Commutation* for *Tithes*, by Mr. BENJAMIN PRYCE, has been honoured with a public mark of this Society's approbation. The ſubject, with relation to agricultural improvements in this nation, is of the firſt collateral importance. And though it may poſſibly be objected, that any alteration in the preſent ſyſtem of Church eſtabliſhment is not an object of contemplation within the province of an agricultural ſociety, nor a ſubject on which ſuch an aſſociation may be fully qualified to judge; yet is it unqueſtionably a ſubject, on which any body of men, as well as any individual, has a right to form and to give an opinion. The influence which any ſyſtem, for tithing the

produce

produce of *husbandry*, muſt have upon its progreſs, will be ever proportioned to the wiſdom or imperfection of the ſyſtem ſo eſtabliſhed. And in general, though the body of farmers ſhould be allowed to be perſonally *intereſted* in ſhifting the burden which lies immediately upon them, it may be alſo fairly allowed that they are the beſt judges of the irkſomeneſs of the mode whereby they are taxed in their labour.

In the preſent age of liberal enquiry, which has reached the minds of intelligent farmers, as well as other citizens of the realm, there are not wanting many among them, who fully diſcriminate between the preſent legal rights of the clergy, and the erroneous ideas of religion and policy which gave riſe to the peculiarities of thoſe rights. They are convinced, in common with all men of ſenſe and reflection, that while it is their preſent duty to pay the clergy their due, according to the legal eſtabliſhment, and to give them as little trouble as poſſible in the payment; yet that they are warranted in *complaining* of a ſyſtem, which both in its nature and tendency is unequal, inconvenient, and vexatious.—Hence it is not unnatural for ſuch a Society as this to countenance a diſcuſſion of ſuch a topic, and to conſider itſelf as acting the part of a *common friend* to the Clergy and the Laity, by endeavouring to point out a mode of ſupport for the former, more equal and righteous among themſelves, more compatible with the principles of peace, more promotive of univerſal improvement, and, which is eſpecially deſirable, leſs invidious to the claimant.

Such were the views of this Society in offering a premium for the beſt-written paper on the ſubject of a Commutation for Tithes: and though the Society is not ſo ſanguine as to expect, that through their means a national tithe revolution ſhall be brought about; yet have they a hope, that, by agitating the ſubject, the attention of ingenious men, both in and out of parliament, may be the more excited

cited to attempt ſome poſſible improvement, where improvement is ſo much to be wiſhed.

The two next following papers, from the ingenious Mr. WIMPEY, reſpecting the culture of *Turnips*, the management of the *Dairy*, &c. are inſerted as practical treatiſes, on ſubjects of eſtabliſhed importance; and it is preſumed they will not be found to diminiſh the reputation of their intelligent author. On the former ſubject much has been written in detached publications; and from the importance of the Turnip huſbandry, both as a ſyſtem of cultivating the ſoil, and as an abundant ſource of food for cattle, too much praiſe cannot be eaſily beſtowed.

It were much to be wiſhed, that the example of Mr. WIMPEY may excite other Gentlemen, equally capable of reflection and experiment, to bend their attention to the properties and management of this delicate plant; till, if poſſible, ſuch a knowledge of it may be obtained, as to enable the farmer to guard it more effectually from its grand enemy the FLY; by the ravages of which ſo much real calamity is frequently produced. Of the latter ſubject, treated by Mr. WIMPEY, it would be needleſs to expatiate on its general utility. The enormous advance of the prices of butter and cheeſe, eſpecially the latter, within the laſt very few years, has rendered an enquiry into the general conduct of the Dairy, and of Dairy farming, particularly commendable; and it muſt give the Society and the Public great pleaſure to find, that men of Mr. WIMPEY's abilities and information turn their attention to a department of rural œconomy, ſo much connected with the comfortable ſupply of the rich and the poor man's table.

The extracts of letters from Sir THOMAS BEEVOR are intereſting in their kind, as might be expected from the elegant pen of ſo accurate an experimentaliſt; and it is not without reaſon that we hope to receive from Sir THOMAS, as well as from ſeveral other Gentlemen,

tlemen, another year, ſuch accounts of the nature and effects of the *Mangel Wurzel*, as a vegetable for the table, and a food for different kinds of cattle, as may determine the value of this new and celebrated exotic.*

To Dr. Fothergill, and Mr. Hayes, for their laudable attention to the cultivation of *Rhubarb*; and to the latter Gentleman for his other communications, the Society, on the public behalf, has been laid under new obligations.

The remarks on *Fruit Trees*, from Mr. Grimwood, Mr. Samuel, Mr. Gillingwater, and our ſcientific correſpondent Mr. Gullet, will be found intereſting, not only to the lovers of horticulture in particular, but to our country readers in general.

* It is with pleaſure the Society embraces this opportunity of paying a full tribute of reſpect to Dr. Lettsom, (though not an immediate correſpondent on the ſubject) for that ſteady and laudable zeal which he has ſhewn for the introduction of ſo promiſing a root into this country. By ſuch generous exertions, whenever well directed, an individual may ſecure, even againſt the force of prejudice, the moſt laſting advantages to his fellow-citizens, and to poſterity.

To

To the experienced and accurate Mr. ONLEY, of *Stifted-hall* in Eſſex, the Society owes many thanks for his preſent and paſt favours, and ſolicits the continuance of his correſpondence. When Gentlemen of ſuch practical knowledge, and exact obſervation, are diſpoſed to favour inſtitutions of this kind with their ſupport, we may hope to gratify the public attention with a beneficial and laſting effect.

To Sir JOHN ANSTRUTHER, the Society continues to be indebted for careful records of experiments, in a branch of huſbandry growing into general eſteem, and which can never be better recommended than by repeated experiment, calculation, and compariſon.

To Mr. NEHEMIAH BARTLEY, no ſmall ſhare of commendation is due, as well for the variety of his uſeful experiments, as for the candour and liberality with which he imparts them.

The

The correſpondence of Miſs Henrietta Rhodes, whoſe attentive obſervation and judicious opinions do equal credit to herſelf and her ſex, the public cannot be wanting to value; and this Society wiſhes her continued remarks on the curious ſubject of her particular attention.

Mr. Winter's own publication has anticipated what we might otherwiſe have had pleaſure in communicating from his pen; but too much credit cannot be given him for his practical exertions, and endeavours to promote the *Drill* huſbandry. His drilling machine, of which we give a repreſentation by his own plate, promiſes much utility on lands favourable to ſuch a mode of cropping; and the public has great reaſon to expect the gradual advancement of this mode, through different parts of the kingdom, to the great increaſe and perfection of the different ſpecies of grain, as well as the ſaving of prodigious quantities of ſeed, heretofore loſt to the nation.

The Reverend Mr. COOK, (a deſcription of whoſe drilling machine was given in our laſt volume) having furniſhed for this publication an ample account of the ſucceſs attending the uſe of that machine; we inſert the ſame on a principle of impartial juſtice to him, and of unbiaſſed attention to the public advantage.

To Mr. WOODBINE, Mr. WAGSTAFFE, Mr. CROCKER, Mr. HAZARD, Mr. WEBB, Mr. POTTICARY, Mr. ANDREWS, Mr. KIRKPATRICK, Mr. TRIFFRY, Mr. BAKER, and various other correſpondents whoſe names do not appear, the Society owes a return of acknowledgements for their obliging miſcellaneous communications;—ſuch acknowledgments are cordially returned, whatever reaſons may have induced an omiſſion of ſome articles, both well intended and reſpectably written.

Laſtly, we have to acknowledge very particular obligations to Dr. FALCONER, for the excellent Eſſay with which this volume concludes. The co-incidence of the Doctor's deſign

ſign with the public views of this Society, is ſufficiently obvious. The preſervation of a claſs of men who are the hands and ſinews of national ſtrength, is a firſt object of ſound policy, as well as of genuine benevolence. If that praiſe be juſt, which has been fully beſtowed on a ſentiment of GOLDSMITH,

"But a bold peaſantry, their country's pride,
"If once deſtroyed, can never be ſupplied,"

we cannot be too ſtudious of preſerving their health and vigour. And while many of our readers will be pleaſed and inſtructed by the Doctor's judicious counſel, his piece will be conſidered as highly worthy the adoption of the Society, and worthy himſelf as one of its original founders.

To conclude. Though the Society cannot undertake to vouch for the perfect accuracy of every account, nor for the juſtice of every opinion contained in the maſs of materials, which from time to time we may be able to lay before the public; yet may we

reaſonably

reaſonably hope, that the continued communications of our correſpondents, on various important matters, will be found an accumulation of ſcience either immediately obvious, or remotely tending to national good. And if, in ſome unavoidable inſtances, an inequality of intrinſic knowledge ſhould be diſcovered, it is but the inequality of human endeavours, ever to be expected.

Thoſe Gentlemen, who, from their own practice and ſagacity, are the beſt qualified to diſcriminate between truth and error, will ever be found the moſt diſpoſed to candour. And while every man deſirous of improvement in rural purſuits will make trial of a new proceſs with caution; any new acquiſitions of agricultural ſkill, which may take their riſe even from the ſuggeſtions of theory, will be deſervedly conſidered as ſo many teſtimonials to the public benefit of eſtabliſhments like this.

BATH, *May* 1, 1788.

LETTERS

LETTERS

TO THE

BATH AND WEST OF ENGLAND AGRICULTURE SOCIETY.

To the SECRETARY.

SIR,

I Obſerve with pleaſure that the Society inſtituted at Bath, for the encouragement of Agriculture, Arts, Manufactures, and Commerce, continues to publiſh many valuable letters and papers communicated by gentlemen in every part of this iſland. And as I am diſpoſed to believe that theſe papers are very generally read, I have been induced to communicate to you the reſult of a few experiments I have made on the culture of Potatoes, which contain ſome diſcoveries that ſuch friends as I have imparted them to think of great importance. If the Gentlemen of your Society ſhall view them in the ſame light, and think them worthy a place in their next volume, theſe papers are much at their ſervice.

Having attended very particularly to the ſubject of agriculture for many years paſt, I have obſerved with not leſs concern than amazement, the ſmall advances that have been made in this uſeful art, when compared with that of other arts of leſs general utility, and have endeavoured to inveſtigate the cauſe of this phenomenon. I find it leſs difficult to diſcover the cauſe of this ſtationary ſtate of our knowledge, than it is to remove the obſtructions that ſtand in the way. Without entering here upon the queſtion at large, I ſhall content myſelf with obſerving, that the length of time neceſſary for making an experiment in agriculture, and the difficulty of diſcovering all the circumſtances that may vary its reſult, are among the chief cauſes of the ſmall progreſs that has been made in this uſeful and neceſſary art. Man, impatient of delay, and anxious to get forward, becomes tired of the ſnail-like progreſs he muſt make if he were to ſubmit all his facts to the teſt of *experiment*. To avoid that tireſome progreſs, men in general have been willing to admit *experience* as a mode of acquiring knowledge on this ſubject, ſufficiently accurate for all the purpoſes of life. Among practical farmers this is ſo much the caſe, that they rely entirely on *experience* as an infallible guide, and condemn *experiments* as abſurd abfurd and unneceffary. By this means the fubject is only imperfectly inveftigated, and uncertainty pervades every department of it.

In compliance with common cuftom, by *experience* I here mean thofe general obfervations, collected from an extenfive courfe of practice, which, by frequently recurring, have made a deep and lafting impreffion on the mind; and by *experiment*, I mean thofe fpecial trials that have been made to afcertain particular facts with accuracy. In the firft cafe, facts are admitted as proved by the frequency of their recurrence, and the fuppofed notoriety of their correfpondence with each other, without being fubjected to any other criterion of accuracy but a general recollection of their frequency and univerfality. In the laft cafe, like mathematical truths, nothing is admitted till it be clearly proved. Upon a fair inveftigation it will appear that a practical farmer, in different circumftances, muft fometimes place reliance on the one, and fometimes on the other of thefe two modes of acquiring knowledge; and that without the aid of both, he never can underftand his bufinefs compleatly.

The bufinefs of a practical farmer naturally divides itfelf into two branches. *One* that embraces the œconomical detail of the operations of hufbandry; *the other* relates to that degree of fcientifick knowledge which directs to thofe operations that ought to be performed. In the firft fenfe he may be compared to a *mafon*, in the laft to an *architect* of a new building. Now, though it fhould be admitted, that, by a conftant courfe of attentive *experience*, a man may in time acquire fuch a knowledge of the detail of the practical operations of hufbandry as could not otherwife be obtained, and might thus come to know, by a fort of mechanical habitude, without much forethought or reflection, the various obftructions that ufually occur in the courfe of bufinefs, and the eafieft means of furmounting them;—though he may come to know in what manner to conduct his different operations, fo as not to interfere with, or to interrupt one another, and be thus able to make fuch ufe of time, as that none of it be either mifapplied or loft by the different perfons he has occafion to employ;—yet all thefe allude only to the firft department of bufinefs, which, of whatever importance it may be to the fuccefs of thofe who follow the bufinefs of agriculture for a fubfiftence, is only a part of that profeffion; and much knowledge remains to be acquired in the other department of agriculture, with regard to which

which *experience* would ſerve but as a very imperfect inſtructor.

It is indeed impoſſible for any man who practiſes agriculture to avoid obſerving, that better or worſe crops may be obtained from the ſame field in different circumſtances, and that certain ſoils are better adapted to yield good crops of one kind of produce than of others. It is as impoſſible for a man, whoſe ſubſiſtence depends upon the produce of his fields, to avoid forming conjectures as to the cauſes of theſe diverſities; and in the courſe of a long and attentive obſervation it muſt probably happen, that ſome of theſe *conjectures* may be right. But as this judgment is formed merely from a complex view *of the whole*, in which a great variety of particulars are blended indiſcriminately together, it is impoſſible for the mind to diſtinguiſh in this way, with any degree of certainty, thoſe circumſtances that are of *eſſential* from thoſe that are of *trivial* importance. The imagination is thus left at full freedom to exert its influence; and ill-grounded theories ſo warp the mind as to make it believe that it ſees certain facts as clearly proved, which are nothing elſe but a ſpecious deluſion. Nor is it poſſible ever to correct the falſe judgments that are thus formed, but by calling in the aid of experiment; which, by carefully ſeparating every circumstance that can in any case affect the result, and viewing it distinct and apart from all others, gives full room to perceive what degree of weight it ought to have in every practical case, and to shew how far it is either essential or unimportant.

The experiments which accompany this letter, and the observations that result from them, sufficiently evince the justness of these remarks, and afford a very convincing proof of the necessity of subjecting the different cases that daily occur in agriculture to the test of accurate experiment, if ever we hope to obtain such a thorough knowledge of facts as to introduce that degree of steadiness in the practice of husbandry, of the want of which we have still reason so justly to complain.

I hope any apology for the trouble I now give you will be unnecessary. I remain, with great respect for the worthy members of your Society,

Sir,

Your most obedient, and

most humble servant,

Cotfield, near Leith,
Nov. 28, 1786.

JAMES ANDERSON.

Miſcellaneous Experiments and Obſervations on *the Culture of Potatoes, and ſome other Plants; written originally in the year* 1778, *with ſome additional Remarks of a later date. By* JAMES ANDERSON, L. L. D. F. R. S. *and* F.S.A. *Scot.*

PAPER FIRST.

ON THE NATURE OF THE SEEDS MOST PROPER FOR BEING PLANTED.

THE Potatoe has been cultivated in Britain for half a century paſt, with great advantage to the community; but many particulars reſpecting its culture are ſtill involved in uncertainty. To point out the means by which that uncertainty may, in ſome caſes, be removed, is the chief deſign of the following eſſay.

§. I.

Among other particulars, it ſtill remains a doubt with practical farmers, whether it is moſt profitable to uſe ſmall potatoes *uncut* for ſeed, or large ones *cut into pieces*. This does not indeed appear to be a matter of *doubt* with any one individual, if he alone ſhould be conſulted on this head; but the uncertainty appears when many are conſulted. Every one is ready to decide poſitively in ſavour of one or other of theſe modes of practice; but when the votes are collected, it is found that they are nearly equally divided between the two; and when enquiry is made into the reaſons on which theſe oppoſite opinions are founded, it appears that they reſt upon no better foundation than theoretical conjectures: for I have never been able to learn, upon the moſt attentive enquiry, that a ſingle comparative trial has been made with a view to aſcertain this fact. The culture of this plant never attracted my own particular attention till lately, and therefore I never did think of aſcertaining this fact by experiment till the year 1776, at which time the following comparative trial was made with that intention.

Experiment First.

April 26, 1776, Four rows of potatoes were planted in a piece of garden ground, without dung, for the ſake of the experiment; there being no variation of ſoil in any part of the experiment ground. Theſe four experimental rows were planted contiguous to one another, and at equal diſtances; other potatoes were planted on each ſide of them, at the ſame diſtance as they were from each other, to render all these rows as much alike each other as possible in all respects. The four experimental rows consisted of plants of the following kinds:

1st row. Small potatoes planted whole.

2d. Potatoes somewhat larger than the former, cut into two equal parts.

3d. Pieces cut from the small end of large potatoes, with one eye in each.

4th. Pieces cut from the large end of the same potatoe, with one eye in each.

To understand the meaning of the last part of the experiment, it is necessary to observe, that the kind of potatoe used in this experiment (and in all the other experiments in this essay, where not otherwise related) was that commonly known here by the name of the *white kidney potatoe.* The bulbs of this kind of potatoe are usually of an oblong shape, flatted a little, having one end considerably smaller than the other: the colour yellowish white, without any tinge of redness. The small end of this potatoe, which is always opposite to the umbilical eye, by which it adheres to the stem, is usually filled with a cluster of bud-eyes, very close upon one another; so that the slices taken from this end, with a single eye in each, are of necessity very small; whereas those that are cut from the opposite end, in which the eyes are placed much thinner, are always of a much larger size.

These potatoes were properly hoed, and kept free from weeds during the summer, and on the 30th of October they were dug up, and, after being properly cleaned, the weight

weight of the produce of the refpective rows was found to be as under, in avoirdupoife weight.

	lb.	oz.		lb.	oz.
1ſt row, - -	18	0	*3d row*, - -	12	5½
2d - -	16	13	*4th* - - -	36	4

The difference between the produce of the third and the fourth rows appeared to me aftonifhing: and as the plants in the fourth row confifted of much larger pieces, and as thofe in the third row were fmaller than any of the others, this experiment feemed to indicate, that the weight of the produce depended in a great meafure on the weight of the feed planted.

It likewife feemed, from this experiment, that whole potatoes might in fome cafes be more profitable for feed, and in others lefs fo, than cuttings; for the firft row exceeded the fecond and third, though it fell greatly fhort of the fourth. The cuttings in the fourth row were much larger, and thofe in the third much fmaller than the whole potatoes in the firft.

It deferves to be remarked, that the vigour of the ftems of each of thefe rows was nearly in proportion to the weight of produce above ftated.

It is alfo of importance to remark, that although the foil, at the time of planting, was in every refpect equal through the whole of the experiment ground; yet, at the time of taking up the plants, that part on which the fourth row grew, was in much better order, and feemed to be much richer, than that where the others had grown, efpecially the third.

It

It is likewise worth noting, that a row of the potatoes which grew beside these, having been taken up by itself, and the bulbs cleaned, was found to weigh 23 pounds. These were from seed cut in the ordinary random way.

§. II.

As the foregoing experiment seemed to point towards an important discovery with regard to the culture of this valuable plant, I resolved to repeat it next season with still greater accuracy, which was accordingly done as under.

Experiment Second.

In the month of April, 1777, a piece of ground was prepared for the experiment. This had been in grass some years, and now got a slight kind of trenching barely to cover the sward, without any dung. It was found that this small piece of ground could contain exactly twenty plants in length, at sixteen inches from each other; and it was divided into rows crossing these at right angles, at the distance of sixteen inches from each other also; so that the plants stood in square sixteen inches from one another, in every direction. The soil of this patch was thin and poor, insomuch, that when in grass, the crop was so scanty as scarcely to admit of being cut with the scythe; but no dung was put upon it, on account of the difficulty of spreading it so equally as not to affect the accuracy of the experiment.

On the 5th of May, twenty plants of each of the following kinds were provided and planted, each kind by itself, in a single row; all the plants in each row being, as nearly as possible, of one size. A row of potatoes cut promiscuously

miſcuouſly having been firſt planted next the edge of the plot for the ſake of accuracy.

	ounces
1ſt row. Small potatoes whole. The twenty plants together weighed - -	5½
2d. Small potatoes cut in two - - -	3½
3d. Small pieces cut from the ſmall end of large potatoes, with one eye in each - -	1⅓
4th. Pieces of an equal ſize with the former, cut out of the large end of large potatoes, with one eye in each - - -	1½

[N. B. *Though it was not expected that any difference could ariſe from the difference of circumſtances here noted, yet as this had never been aſcertained by experiment, the fact was not certainly eſtabliſhed. This trial was meant to give it the certainty wanted.*]

	ounces.
5th. Large pieces cut from the great end of the ſame potatoes that were employed in No. 3 and 4, having only one eye in each, -	26¾
6th. Large potatoes, from which all the eyes had been cut out, ſave one about the middle part of the bulb - - -	121½
7th. Large potatoes with one eye only, left in the ſmall end of the bulb, - -	123½
8th. Large potatoes planted whole, as nearly as could be got, of an equal ſize with the former, - - -	124½

[N. B. *No. 6 and 7 were intended to diſcover whether the produce continued to increaſe with the weight of the ſeed planted. The leaving only one eye was intended*

to

to make these plants resemble, as nearly as possible, those in No. 5. The variation between No. 6 and 7 was with the same view as that in No. 3 and 4. No. 8 was intended to discover if plants are damaged in any respect for seed merely by being wounded, and what is the result of planting seeds with many or few eyes.]

EXPERIMENT THIRD.

On the same patch of ground that was prepared for the foregoing experiment, and immediately contiguous to the 8th row in the preceding experiment, (one row only intervening, which will be afterwards taken notice of) was planted on the same day with them, *seven* other rows of seeds, being each of them exactly of the same size and weight with the foregoing; so that it was an exact repetition of the same experiment, intended to save time. The only difference between them was, that the seventh row was here entirely omitted for want of room. The general result of these two experiments was as under; the uppermost row of figures, where double, denoting the result of experiment 2d, and the undermost of experiment 3d.

No. of rows.	No. of seeds that germinated.	Weight of seed. lb.	oz.	No. of roots produced.	Average	Weight of the produce of each row. lb.	oz.	Average weight of the produce. lb.	oz.
1st,	19 20		5½	122 125	123½	6 8	0 12	7	6
2d,	19 18		3½	107 131	119	5 6	13 0	5	15½
3d,	17 15		1½	62 54	58 } 64½	2 1	8 15	2	3½ } 2 6½
4th,	17 17		1½	56 86	71 } 64½	2 3	3½ 0	2	9½ } 2 6½
5th,	20 19		26	190 192	191			12	2½
6th,	20 19	7	10½	315 258	286½	19 16	3 15½	18	1½
7th,	20	7	11½	374	374	18	10½	18	10½
8th,	20 26	7	12½	470 330	400	21 20	5½ 3½	20	12½

From

From theſe two experiments thus carefully collected it appears, that there is ſuch a near coincidence between the produce of the correſponding rows in each experiment, as gives us reaſon to believe, that the average obtained from each row is nearly what would reſult in general practice from planting ſeeds, correſponding to thoſe planted in each of theſe rows reſpectively; ſo that the corollaries deducible from thence may be admitted as general rules in practice.

§. III.

It may, in the firſt place, be inferred, by a careful review of theſe two experiments, *that the produce is not materially affected by planting for ſeed, either whole potatoes or cuttings, or large or ſmall potatoes merely as ſuch; for that it is only incidentally that either of theſe particulars can affect the crop.* The whole potatoes in the firſt row yielded a ſmaller produce than the *cuttings* in the 6th row. Seed from *ſmall* potatoes yielded a ſmaller produce than was obtained from *large* ones, in the 5th, 6th, 7th, and 8th rows; but it yielded a greater produce than was obtained from *the ſame* large potatoes, in the 3d and 4th rows. It ſeems, in the *ſecond* place, to be a fact confirmed by every ſtep in both theſe experiments, *that the weight of the crop is always in ſome meaſure influenced by the* WEIGHT *of the ſeeds planted.* The third and fourth rows, in which the ſeeds were *lighteſt*, yielded the pooreſt crop; and a progreſſion from lighter to more weighty, is obſerv-

able

able in the *produce*, as well as the seeds through the 1st, 2d, 5th, 6th, 7th and 8th rows. Some trivial variations do not disturb the general rule, which seems to be sufficiently established by the general result of the first six rows.

§. IV.

Of all the experiments in agriculture that I have ever seen recorded, that in question exhibits the most interesting result, whether we consider it with respect to the principle from which the phenomena originate, or its great importance in agriculture as a *practical* art. In this last respect, indeed, it promises to be of the utmost utility, because by it we are taught, that without any alteration in the soil or culture, but merely in consequence of a proper attention to the state of the seed to be planted, a crop *nine* times as weighty may be obtained in one case as in another. Is it not astonishing, that a circumstance of such amazing influence should not have been discovered by accident long before this time? And does not this afford a most convincing proof of the necessity of subjecting the common modes of husbandry to the test of actual experiment, in order to obtain a rational degree of certainty, instead of those conjectural opinions that individuals are so apt to rely upon with unsuspecting confidence?

That

That the nature of the ſubſtance from which a plant is to be produced, ſhould have *ſome* influence on the future vigour of that plant, ſeems not unreaſonable to ſuppoſe; yet I believe that even the warmeſt imagination could hardly induce one to ſuſpect *a priori*, that ſuch an extraordinary degree of vigour could be communicated, merely by an increaſe in the *quantity of matter* contained in the ſeed. To me, this circumſtance appeared the more ſurpriſing, as the reſult was extremely different from what I had found by ſome former experiments was produced by plump and lean grain employed as ſeed. The experiment was as follows:

Experiment Fourth.

With a view to know of what conſequence it was in the practice of agriculture, to employ plump or lean grain for ſeed, I planted, April 2d, 1770, upon a ſmall bed of ground in a garden, one hundred of the plumpeſt grains of oats that I could pick out from a large parcel of *unmixed* oats, in five rows, five inches row from row, and one inch between each plant in the rows. On another equal ſpot in the ſame ground, I planted at the ſame time, and in the ſame manner, one hundred of the hungrieſt grains I could pick out from the ſame parcel of oats: but to inſure againſt contingencies, I alſo took as many of the ſmall hungry grains as equalled *in weight* the hundred plump grains abovementioned, which, when numbered, I found amounted to one hundred and ſeventy. Theſe 170 grains I planted in five rows, each of the ſame length as the former, and diſtant from each other five inches, ſo that the hundred and ſeventy

ſeventy bad grains occupied preciſely as much ground as the hundred good grains.

RESULT.

No. 1ſt. That diviſion on which a hundred good grains were ſown, produced ninety-five plants.

2d. That on which a hundred lean grains were ſown yielded ninety-ſix.

3d. That diviſion on which was ſown the one hundred and ſeventy hungry grains, yielded alſo ninety-ſix.

On the firſt appearance of the ſeed leaves above ground, thoſe of No. 1ſt were broader, and more ſucculent than thoſe of the other two plots; but as the plants advanced towards perfection, the difference in appearance gradually began to diſappear, and long before harveſt it was not poſſible to remark any difference in the healthineſs and luxuriancy of the ſtalks in any of the three diviſions. The grain when ripe was equally healthy in No. 3d as in No. 1ſt, and the crop ſeemingly as weighty in every reſpect: but this I could not aſcertain with the certainty I wiſhed, on account of the deſtruction by birds.

The reſult of this experiment was, in truth, very contrary to what I had expected. If No. 2d only had been ſown with the lean grain, I ſhould have attributed the health and vigour of the plants to its thinneſs: but, without the aid of that circumſtance, the plants in No. 3d were equally ſtrong and vigorous. I mean not, however, at preſent, to make any farther uſe of this experiment than barely to remark how very dangerous it is in farming, to rely implicitly on reaſoning from analogy between two caſes that are not *in every reſpect* alike, though they may reſemble each other in many ſtriking particulars. It would not, for example, ſeem very unnatural for a perſon who had made only *one* of theſe experiments, to conclude *from analogy*, that the reſult in the ſomewhat ſimilar caſe, which he had not tried, would be ſimilar to that which he had tried: yet it appears, that with regard to grain (that kind of it at leaſt which had been proved) a difference in the *weight* of ſeed, if it has *any* effect on the future crop *at all*, is ſo little as ſcarcely to be perceptible; whereas, with reſpect to the plants of potatoes, it is ſo great as to augment or diminiſh the total amount of the crop in the *ratio* of nine to one. This, at the ſame time that it ſhould teach the farmer to be extremely cautious how he ſuffers his mind to be influenced by vague reaſoning, ought ſtrongly to incite him to redouble his attention, and by well-choſen experiments endeavour to obtain ſome kind of certainty in the knowledge of many particulars, wherever he finds that his opinions have been adopted in conſequence of early prejudices, or crude indigeſted notions ariſing from theories that have not been ſufficiently underſtood.

§. V.

But although it appears, from experiments *firſt*, *ſecond*, and *third*, ſufficiently obvious, that the crop of potatoes is augmented by the weight of ſeed, yet it alſo appears from experiments *ſecond* and *third*, that the weight of produce is not augmented *in the ſame proportion* with the weight of the ſeed: for although the weightieſt ſeeds have always yielded the weightieſt crop *in proportion to the extent of ground*, yet the lighteſt ſeeds have as invariably produced the greateſt return *in proportion to the weight of ſeed planted.*

planted. That the reader may be enabled to obſerve every particular relating to theſe two proportions, the following table has been conſtructed. In this table is expreſſed the quantity of ſeed, and the produce of an Engliſh ſtatute acre, proportioned to the weight of ſeed and produce in the different rows of the preceding experiments, together with the returns from the ſeed in each row, and the clear produce after deducting the ſeed.

That thoſe who chooſe it may be able to follow theſe calculations, they need only to be informed, that an acre would contain 24,502 plants at ſixteen inches from each other: all the other data neceſſary are expreſſed above.

Rows correſponding to thoſe of the ſame numbers in Exp. 2d and 3d.	Quantity of ſeed required to plant an acre in the proportion of each row reduced to buſhels and decimals.	Quantity produced from an acre, in the proportion of each row, reduced to buſhels & decimals.	Proportional returns of ſeed from each row.	Clear produce from an acre, in the proportion of each row, after deducting the ſeed.
	Buſh. Dec.	*Buſh. Dec.*		*Buſh. Dec.*
1ſt,	7.50	161.30	21.4	153.80
2d,	5.13	130.5	25.3	125.37
3d, 4th,	2.05	52.6	25.7	50.65
5th,	35.5	266.5	7.5	231.00
6th,	167.4	396.1	2.4	228.7
7th,	168.6	400	2.3	231.4
8th,	170.2	453.9	2.6	283.7

From this table it appears, that the 3d and 4th rows, in which the ſmalleſt quantity of ſeed was planted, yielded the greateſt returns, *in proportion to the ſeed*, but the ſmalleſt *in proportion to the extent of ground.* The returns of ſeed being as 25.7 to one;

one; whereas that of No. 8th was only 2.6 to one. But the total average produce of the 3d and 4th rows was only 52.6 buſhels; whereas that of the eighth row was 453.9 buſhels.*

To obtain a juſt notion, however, of the profit that would be derived from cultivating a field in the one or the other of theſe ways, it is neceſſary to deduct the ſeed in both caſes from the groſs produce, the remainder only denoting the free produce. The laſt column in the table above marks this free produce, in all the different caſes above ſtated. And from this table it appears, that the total free produce from the ſmalleſt ſeed here employed was only 50.65 buſhels per acre; and that where the largeſt ſeed was employed, amounted to 283.7, ſo that one acre in the laſt caſe yielded nearly as much free produce as ſix acres in the firſt.

Hence it ſeems reaſonable to infer, that it is in no caſe profitable to plant ſmall potatoes, or ſmall cuttings, unleſs where it is meant to increaſe as faſt as poſſible a favourite kind; in which caſe it may be ſometimes eligible to plant pieces very ſmall, as in that way the kind will be moſt quickly multiplied.†

* By experiments more at large ſince that time, and on a richer dunged ſoil, I have obtained a return from ſeeds even larger than thoſe in No. 8th, in the proportion of at leaſt *ten* to one, ſo that the very ſmall returns in this experiment muſt be aſcribed to the great poverty of the ſoil.

† Since the above was written, I find reaſon to believe, that the returns from *large* potatoes may be augmented greatly beyond what it was in this experiment; whether

§. VI.

By comparing No. 6th and 7th with No. 8th, in experiments ſecond and third, there is ſome room to ſuſpect that the ſeeds may poſſibly have been injured by the wounds they received in having their eyes cut out, as the produce in No. 6th and 7th does not ſeem to be quite ſo great in proportion to the ſeed as in No. 8th. But this difference is not ſo conſiderable as to enable us to ſpeak with any degree of certainty. Had it even been greater than it is, there would ſtill have been room to doubt whether it had been occaſioned merely by wounding the ſeeds, or in part alſo by diminiſhing the number of the eyes. The following experiments would tend to elucidate theſe particulars:

1ſt. Take any determinate number of potatoes, all of one ſort, and of an equal weight each, and having ſeparated them into two equal parts, plant all thoſe of one diviſion *whole*, and let all the plants of the other diviſion before planting, be wounded with a knife in many places, without cutting out any of the eyes. Obſerve the reſult.

2dly. Take, in the ſame manner, another determinate number of potatoes, of the ſame ſort, all of equal weight, and having ſelected an equal number of the ſame kind of potatoes ſomewhat larger each than the former, wound theſe laſt deeply in various places, and cut out from them

whether it could by any peculiarity of culture be brought to equal that from ſmall, my experiments, which have been interrupted by other avocations, do not enable me to ſay—but it is not at all improbable.

feveral deep flices, fo as to reduce them to an equal weight with the former, taking care not to cut out or wound any of the eyes. Plant thefe in equal circumftances, and obferve the refult.

3dly. Repeat the experiment of the 7th and 8th rows of experiment fecond, with proper caution: for I have a fufpicion, that in my experiment the eyes in the plants of the 7th row had not been cut out deep enough, to prevent them from fending forth ftems.

4thly. Take fome large flefhy cuttings, with one eye only in each, all of an equal fize, and having felected an equal number of whole potatoes, equal in weight to thofe cuttings, plant them, and obferve the refult.

[Since the above was written, other avocations have prevented me from repeating thefe, and many other experiments propofed in this effay. To fome the importance of thefe experiments will appear doubtful, and many will feel a ftrong propenfity to foretel what would be the refult, and therefore will think it unneceffary to prove it by actual trial. This prefumptuous propenfity has tended in a wonderful degree to retard the progrefs of agriculture, and cannot be too cautioufly guarded againft. Before we can attempt to make any *decifive* experiments on the beft method of cultivating this plant, fo as to obtain in every fituation the greateft poffible crop that circumftances admit of, all thefe previous queftions muft be fully difcuffed. From the few experiments above recorded, we are enabled to perceive in what manner many hitherto inexplicable peculiarities recorded concerning the culture of this valuable plant may be accounted for, that have been explained far otherwife.

Tho' it does not appear probable that the mere wounding the bulbs will affect the crop, yet it is certainly within the

the bounds of *possibility*, and therefore the fact should be ascertained. As to diminishing the number of eyes, the *probability* that it may affect the crop appears very strong. Every stem which springs from a potatoe becomes in time a distinct plant, which spreads its own roots around, and sends forth its own clusters of potatoes in the same way as if it were a distinct and separate plant. By having many or few of these, therefore, the crop may certainly be affected —but how far no one at present can say; and therefore no one can make an accurate comparative experiment on the culture of potatoes in general.]

§. VII.

There seems to be no reason to suspect that eyes taken from any particular part of the bulb are possessed of a degree of prolifiacy greater than those taken from any other part of it, independant of the *size* of the fleshy part that adheres to the eye. This appears by comparing the 3d with the 4th, and the 7th with the 8th rows in the foregoing experiments.

[It is however highly probable that a difference in the crop, either with respect to the number and size, or general weight of the whole, would result from planting *large* cuttings of equal weight, taken from the big end of large potatoes, or from the point, as many eyes would be in the last in comparison of the first. This is therefore one of the many preparatory experiments that requires to be made.]

§. VIII. Hitherto

§. VIII.

Hitherto I have only taken notice of the total *weight* of the crop; but as the *value* of that crop is, in many cafes, affected by *fize* of the bulbs, it is neceffary to attend to thofe circumftances that may tend to increafe or diminifh their fize. With a view to that particular I have, in the preceding experiments, recorded the *number* of potatoes produced in every cafe, as well as their weight.

It is commonly imagined, that if the feeds planted contain many eyes, the bulbs produced will be numerous, but fmall; and that larger bulbs in fmaller number are produced in plants that have only one, or few eyes: hence it is concluded, that *whole* potatoes planted for feed will always produce a greater number of *fmall* potatoes; and cuttings will yield larger potatoes, though fewer than thofe. It does not, however, appear, that this hypothefis is fupported by the foregoing experiments. In the average table, §. II. we find that the bulbs produced from the 3d and 4th rows, which confifted of plants with one eye only in each, were fmaller, as well as lefs numerous, than thofe in the 1ft and 8th rows, which confifted of plants that contained many eyes. On the other hand it appears, from the fame experiments, that the bulbs produced

produced from the 5th row, in which the ſeeds contained one eye only, were larger than thoſe in the 8th row, (conſiſting of plants with many eyes) in the proportion of 20 to 24 nearly. I would draw no concluſion on this head from the 6th and 7th rows, becauſe, as I have already obſerved, I ſuſpect that ſome of the eyes in theſe had not been cut out deep enough to prevent ſome of them from puſhing out ſtalks; for I obſerved that in theſe two rows, as well as in the 8th, there were many ſtems aroſe from each plant. It deſerves however to be remarked, that ſeveral ſtems ſprang from the roots of the others, and ſome of theſe at a conſiderable diſtance aſunder, although in theſe I think I am certain there was only *one* eye in each cutting, for I cut them all with my own hand, and was as careful as poſſible to examine them with attention; ſo that nothing *certain* can be inferred from the number of ſtalks that ſprung from one plant.

§. IX.

It is commonly imagined that the ſize of the bulbs is augmented, and their number retrenched, by cutting off the ſupernumerary ſtalks at the firſt hoeing, ſo as to leave only one ſtem at each plant; but I never heard of any experiment by which this fact has been aſcertained. Therefore in equal cir-

cumſtances

cumſtances plant two rows of the ſame kind of potatoes whole, the plants being all of equal weight; and in the firſt hoeing cut off all the ſtems ſave one to each plant in one row, and in the other leave all the ſtems. Obſerve the reſult.

Plant at the ſame time other two rows in every reſpect as the former, but inſtead of cutting off the ſupernumerary ſtems, *pull* them up by the hand.—Obſerve the reſult.

[Since the above was written, I attempted this experiment, but I found that new ſtems ſprung up from thoſe that were cut over, ſo as rather to augment than diminiſh their number, and alſo to retard the ripening of the ſtems; and as theſe ſtems bleed conſiderably when cut, it ſeems probable that the plant muſt be weakened thereby. But as this experiment was not made with ſufficient accuracy, no *certain* concluſions can be drawn from it.

It appeared to me that if potatoes were planted very ſhallow, more ſtems were always produced than if they were planted deeper; but in no caſe where potatoes are planted whole, does it ſeem that a ſtem is produced from every eye. Perhaps the beſt way of diminiſhing the number of ſtems from large potatoes, is either to let them ſpring before they are planted, or to take them up ſoon after they have germinated, and to rub off the young ſprouts as many as you incline. The germ becomes a plant adhering to the preſent bulb, whoſe roots ſpread on its ſurface before they ſtrike into the ground, and in that ſtate may be eaſily ſeparated, and poſſibly might be employed as plants.

I have

I have yet made no experiments to try if the crop be ſenſibly affected by planting the ſeeds deep or ſhallow, and by conſequence do not know what is the moſt proper depth to plant them at. This particular requires to be elucidated.]

§. X.

Although it appears, from the foregoing experiments, that the weight of the crop was always greateſt where the ſeeds planted were the moſt weighty, yet it would be too raſh in us from thence to infer that as great a crop could *in no caſe* be obtained from the ſame extent of ground, if it were planted with ſmall potatoes or ſmall cuttings, as if it were planted with large ones. For, as it is probable that the more bulky ſeeds would require a greater ſpace to nouriſh them properly than ſmall ones; ſo it is not *impoſſible*, that if theſe ſmall ſeeds were planted at a proportionally ſmaller diſtance, the crop might equal that obtained from the larger ones. Though it does not ſeem *probable* that this would be the caſe, and though it appears probable, were it even ſo, that the practice would be troubleſome and inconvenient, far beyond any benefit that could reſult from it; yet our firſt buſineſs ſhould be to aſcertain how the matter of fact ſtands, and then enquire into the other circumſtances depending on that fact. To do this in a proper manner, a numerous ſet of experiments would be required, ſomewhat upon the following plan.

The firſt ſtep would be to aſcertain what is the diſtance that ought to be allowed between each plant, when the ſeeds are of any given ſize, ſo as to obtain the moſt weighty crop.

FOR EXAMPLE:

Let it be required to aſcertain at what diſtance from *one another* potatoes weighing half a pound each (that is equal to ten pounds of ſeed in our experiment) ſhould be planted, ſo as to inſure the greateſt poſſible crop.

To do this let a plot of ground be made choice of for this experiment, which was of a good quality, and as equal as poſſible in every part. Let this be divided into ſmall ſquares, each of which ſhould be ſufficient to contain exactly one hundred plants, at each of the diſtances it was intended to aſcertain. Theſe ſquares ſhould be divided from each other by a ſingle row of potatoes planted at one foot diſtance from each other in the rows; and each of the ſquares ſhould be ſo divided as to allow every plant in the outſide rows to ſtand at the ſame diſtance from the diviſion rows, as from the other rows in the ſquare. That is to ſay, each ſquare ſhould be divided into *eleven* equal parts, on every ſide, ſo as to allow *ten* free rows every way, as in the following DIAGRAM, on which the ſmall dots repreſent the diviſion rows, and the larger dots the hundred experimental plants.

DIAGRAM.

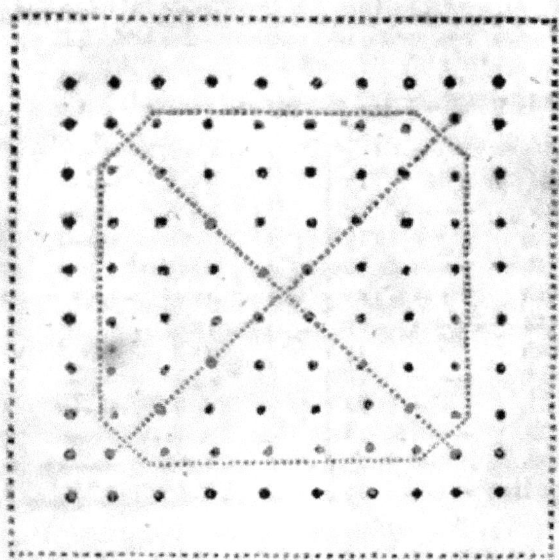

If the whole 100 plants were taken up and weighed, it is probable the experiment would be the more accurate; but ſhould that trouble be thought too great, the ſixty plants contained within the inner line of ſmall dots would anſwer perfectly well; or ſhould that be thought too many, ſtill the 16 plants in the diagonals, with four near the centre marked alſo with dots, ſo as to make in all twenty, would anſwer the purpoſe ſufficiently. All that is here required is, that a certain order of ſelection ſhould be previouſly adopted, and moſt ſtrictly adhered to; for ſhould a random ſelection of a certain proportion of the plants be permitted, this might be done in a particular manner either through prejudice or favour, which might affect the accuracy of the experiment.—A ſquare ſpot divided in this manner ſhould be ſet apart for each of the undermentioned diſtances be-

tween

tween the plants, beginning at twelve inches; as that is ſurely as little as ever could be judged neceſſary for plants of the ſize here ſpecified. The diſtance between the plants in the other ſquares to increaſe as in the table below.

Plants weighing ten pounds per ſcore.

In the 1ſt ſquare to be placed at		12	In the 14th ſquare to be at		25
2d dittto at		13	15th	——	26
3d	——	14	16th	—	27
4th	—	15	17th	——	28
5th	——	16	18th	—	29
6th	—	17	19th	——	30
7th	——	18	20th	—	31
8th	—	19	21ſt	——	32
9th	——	20	22d	—	33
10th	—	21	23d	——	34
11th	——	22	24th	—	35
12th	—	23	25th	——	36
13th	——	24	Which it is imagined will be the greateſt diſtance that need be tried.		
INCHES APART.			INCHES APART.		

Let all theſe plants be carefully hoed and attended to, and when the plants have attained a perfect maturity, let the produce of each ſquare (or of ſuch proportion of it as ſhould be thought proper) be carefully weighed. The reſult would ſhew, with ſome degree of preciſion, what would be the diſtance at which ſeeds of the ſize here ſpecified ought to be planted, ſo as to yield the greateſt crop on a given extent of ground.——But, as it is highly probable that the reſult of this experiment would be different if it were tried on rich and on poor ſoils, it would be proper to have it ſeveral times repeated, trying it on the richeſt and moſt highly manured ſpots, and on others declining from that till they approached to as great a degree of ſterility as this crop could be profitably reared on. Thus would the farmer come to know the proper diſtance at which he ought to plant his potatoes in all caſes. The other experiments that follow in this section, ought to be varied in the same manner.

And as it is also possible that potatoes of different sorts may require a different distance between them, even where the plants are of the same weight, it should be always understood that an experiment of this kind is only to be absolutely relied on when applied to the particular kind of potatoe that was actually tried; therefore, if any experiments of this nature are recorded, the kind of potatoe that was tried should be particularly specified. After this, it is scarce necessary to add, that no intermixture of kinds should be admitted in any of these experiments.

We ought, in the second place, to endeavour in the same manner to ascertain what is the most profitable distance at which plants of all different sizes should be planted. With this view, a number of plants of equal sizes should be selected and arranged into classes by weight, as in the table below, so that the foregoing experiment can be repeated through all its varieties with each class of plants. We shall make the first class consist of plants of half a pound each, or ten pounds per score, as it will be more convenient to weigh the plants by scores than separately; smaller divisions being thus more obviously perceptible than if the plants were weighed individually. The weight of the other numbers is marked in the second columns.

Plants weighing per score.	*lb. avoird.*	*Plants weighing per score.*	*lb. avoird.*
Class 1st	— 10	Class 6th	— 5
2d	— 9	7th	— 4
3d	— 8	8th	— 3
4th	— 7	9th	— 2
5th	— 6	10th	— 1

In

In all these classes the distances should be the same as above, viz. from 12 to 36 inches, varying in each square one inch. In those that follow the greatest distance need not exceed 24 inches, and the smallest distance should be as low as six inches;

Plants weighing per score.		oz.	Plants weighing per score.		oz.
Class 11th	—	14	Class 15th	—	6
12th	—	12	16th	—	4
13th	—	10	17th	—	2
14th	—	8			

Which we will suppose the minimum:—perhaps all below eight ounces might have been omitted, without any detriment to the practice of agriculture. But no harm can ever accrue from ascertaining with accuracy any number of facts in agriculture.

This has the appearance of being a very formidable set of experiments; and it would, no doubt, require a good deal of trouble, and some expence, to execute it properly; so that it should fall to the share of some of those gentlemen of opulence and high rank, who take delight in the study of agriculture. Perhaps few experiments that could be named, would be productive of greater national benefits than that which is here proposed: nor would the expence to a man in easy circumstances be an object of great consequence. Somewhat less than five acres of ground would be sufficient to execute the whole set of experiments *once over*, so as to ascertain with some tolerable accuracy the most advantageous distance for planting each size of seeds, on one class of soils, considered as to their richness, and lead to many probable conclusions as to other soils, which would be of the most extensive benefit in general practice; and the crop obtained would probably repay the greatest

part,

part, if not the whole of the expence. Were ſuch a ſet of experiments carefully made, and properly publiſhed, it would probably advantage the publick many *millions* a year. How much is it to be regretted that a national experimental farm is not ſet apart for making ſuch experiments in agriculture, as it does not beſit practical farmers to make at their own expence!

[In practice at preſent, ſome perſons chooſe to plant large, and others only very ſmall cuttings or little potatoes, yet every man invariably plants them at one diſtance in all caſes, planting the ſmall ſeeds as wide as the large, when he chances to have them of different ſizes. This being the caſe, it ſeems impoſſible, if the foregoing experiments can be relied on, (and I have found by many trials they certainly may) but that the largeſt crop muſt always be obtained from that field which has been planted with the largeſt ſeeds—other circumſtances being nearly alike. And as the variation ariſing from this hitherto unobſerved peculiarity may be extremely great, may we not reaſonably conclude that ſome of thoſe extraordinary variations in the produce of potatoes, which have been remarke d, but not accounted for in any probable manner, may have ariſen ſolely from this circumſtance?]

§. XI.

To enable individuals to eſtimate without much trouble the amount of any crop of potatoes they wiſh to examine, I here ſubjoin a table, ſhewing the number of plants that would be contained in an acre at each of the forementioned diſtances, and the weight of produce from twenty plants in each

 caſe,

caſe, when the whole produce of an acre would be twenty-five, fifty, ſeventy-five. One, two, three, four, five, ſix, ſeven, eight, nine, ten, eleven, or twelve hundred buſhels of 56 pounds each—beginning with ſix inches and ending with thirty-ſix inches diſtance from plant to plant:

That is to ſay,

When the plants ſtand at ſix inches from each other every way, an acre contains 174,240 plants.

If the crop is equal to 100 buſhels per acre, the produce of twenty ſtems would be 0.631 pounds and decimals:

If the crop was equal to 500 buſhels per acre, the produce of twenty plants would be 3.15 pounds and decimals:

And if 100 buſhels per acre, the produce of 20 plants would be 6.31 pounds, as in the firſt line of the table.

And after the ſame manner all the others are to be read in the following table:

Diſtance

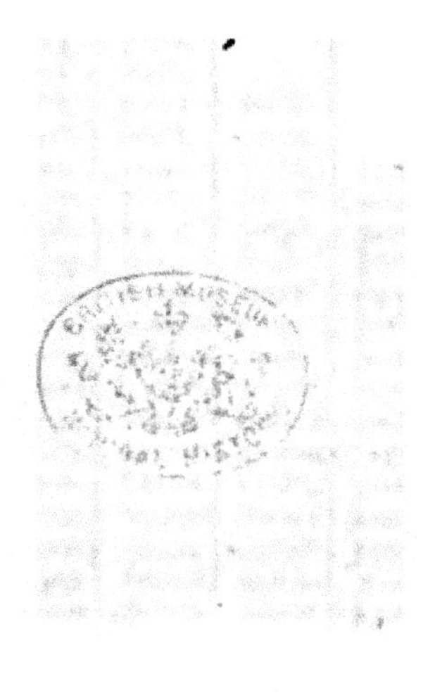

Distance of plants from each other,	No. of plants on an acre.	Th[e]amounted to		
		25	1,100	1,200
		Bushels. lb. dec.	Bushels. lb. dec.	Bushels. lb. dec.
6	174,240	0.152	6.94	7.56
7	128,013	0.218	9.61	10.48
8	98,010	0.285	12.56	13.70
9	77,440	0.361	15.90	17.34
10	62,726	0.446	18.63	21.42
11	51,840	0.540	23.76	2.592
12	43,560	0.642	28.28	30.84
13	37,116	0.754	33.18	36.20
14	32,003	0.877	38.59	42.10
15	27,878	1.005	42.23	48.24
16	24,502	1.142	50.28	54.84
17	21,704	1.292	56.80	61.96
18	19,360	1.445	63.60	69.38
19	17,375	1.611	71.10	77.34
20	15,681	1.782	78.43	85.56
21	14,223	1.968	86.61	94.48
22	12,960	2.160	95.07	103.70
23	11,857	2.380	104.73	114.24
24	10,890	2.571	113.12	123.46
25	10,036	2.789	122.74	133.90
26	9,279	3.017	132.77	144.84
27	8,604	3.254	143.18	156.20
28	8,000	3.500	154.00	168.00
29	7,458	3.754	165.18	180.20
30	6,969	4.017	176.77	192.84
31	6,527	4.289	188.73	205.88
32	6,125	4.571	191.13	219.42
33	5,760	4.861	213.88	233.32
34	5,426	5.155	226.84	247.46
35	5,120	5.454	239.97	261.78
36	4,840	5.841	257.03	280.40

By the help of the ments, or in any other field where the ould be selected for the average, it is ppose a *hundred*: multiply by 5, for fiv inches distance 100 plants were selec2.855, and so of others. In like mat be got by subtracting the lesser frme as low as 20 bushels. Example : find the nearest number to that to bre, which being about the third part g 625 per acre. Any other case may

§. XII.

In the foregoing experiments no attention was paid to ascertain any other part of the produce but the weight of the bulbs only; but as it may happen that the weight of the stems, and the quantity of *apples* produced, may, in some cases, be an object of value, it is worth noting that the strength and weight of the stems were in all the foregoing experiments apparently much in the same proportion as the weight of the bulbs; the stalks being invariably stronger where the crop of roots was weighty than where it was light. The produce of *apples*, should these ever be found to be an object of value, (which there is great reason to think will be the case) increases in a yet higher degree than the potatoes themselves, when the seeds planted are very large; when the cuttings are small, scarcely one apple is seen in a field; when they are large plants, the apples are numerous and of great magnitude, hanging in clusters of nine or ten together; so as in some cases I have known them produce at the rate of more than 200 bushels per acre.

[I mean to make some experiments on the uses to which these may be applied; the result of which shall be in due time communicated to the publick.]

PAPER SECOND.

ON THE EFFECTS OF CUTTING THE STEMS OF POTATOES WHILE GROWING, &c.

§. I.

THE ftems of potatoes, if cut while growing, and ufed green, are found to be a wholfome food for cattle and horfes. But though fome farmers maintain that the produce in potatoes is not leffened by having the ftems cut off while they are in a ftate of vigorous vegetation; others as pofitively infift that the crop is effentially injured by that operation. It is proper that this point fhould be afcertained. Probably the crop is hurt if the ftems are cut over before they have attained a certain point of maturity, though it is poffible they may be afterwards cut without doing any effential injury to it.

The following experiments were made in the year 1779, with a view to afcertain the foregoing particulars:—

Experiment Fifth.

With a view to afcertain the weight of green ftems of potatoes at different periods of their growth, NINE ftems of potatoes, being part of three rows, and three plants in each row, were cut over in the middle of the field as reprefented by by the figures o o o, &c. in the following diagram, at the ſeveral periods marked on the right hand, and were found at each cutting to yield the weight of green fodder marked at each of the periods reſpectively.

D I A G R A M.

Period	Weight of green ſtems, lb.	oz.
Aug. 2d*	7	10
Aug. 10th	7	6
Aug. 17th	7	3
Aug. 22d	7	1
Aug. 29th †	7	0
Sept. 5th ‡	6	2

* Auguſt 2d.—At this time the flowers were juſt beginning to open.

† Auguſt 29th—At this time the apples of the white ſort employed in the experiment were well formed. Red potatoes in the ſame field juſt coming into bloom.

‡ September 5th—At this time the ſtems of the white potatoes were beginning to fade. Apples, ſome of them ripe. Red potatoes juſt paſt the bloſſom, and in full verdure.

It appears from this experiment that the green ſtems are weightieſt at the time potatoes come into bloſſom, (in this experiment the weight of an acre of green ſtems was then equal to 12 tons and a half nearly) and that they become gradually lighter, as the crop approaches nearer to maturity.

Cows eat this forage very readily, as do horſes alſo; but it is not in general accounted a very nouriſhing kind of food.—It is eaten moſt readily when in its moſt ſucculent ſtate.

Experiment Sixth.

[To aſcertain the proportional weight of a crop that would be obtained from a field of potatoes of this kind, if taken up at different periods, one ſtem marked *x* in the foregoing diagram was taken up at each of the periods that the nine ſtems in the laſt experiment were cut over, and they were found to produce when weighed and numbered as under reſpectively:

Produce from one ſtem of Potatoes.

	Weight *lb.*	*oz.*	Number.
Auguſt 2d	0	3½	21
Auguſt 10th	0	7	*omitted.*
Auguſt 17th	0	9½	10
Auguſt 22d	0	14½	15
Auguſt 29th	0	13	7
September 5th	1	7	8

From this experiment it would ſeem, that if the whole crop had been taken up on the 2d of Auguſt, it would have yielded no more than three ounces and a half per ſtem, (at the rate of 125 buſhels per acre) and if let ſtand till the 5th of September, it would have yielded twenty-three

per

per ſtem, (868 buſhels per acre) ſo that at the firſt period the crop would have attained only about one-ſeventh part of its whole bulk.

The reader, however, muſt be cautioned not to rely implicitly on this experiment as concluſive, on account of one material impropriety in the mode of conducting it. *One* ſtem only can never be ſuppoſed to afford a fair average of thirty-two thouſand; more eſpecially when it is adverted to, that the ſeeds planted were in this field cut in the uſual random way; ſo that one might have been found by accident much larger than another, and by conſequence would greatly affect the accuracy of the trial.—Had the nine ſtems contained within the ſmall dotted lines in the diagram been taken up at each period, much greater reliance could have been had upon it. We ſhall afterwards find that there is ſome reaſon to conclude, that the reſult of this experiment is not far from the truth: but as many important leſſons to the practical gardener and farmer could be deduced from this experiment, if carefully made, I cannot help recommending it to the attention of the reader, as one of thoſe radical experiments that cannot be too carefully made and adverted to. For were it known with certainty what is the deficiency of weight that in all caſes would accrue from taking up any one kind of potatoe at a particular period of its growth, the practical agriculturiſt could compute with great accuracy whether the additional price he could receive for the produce at an early period, together with the uſe he could make of his ground after it was cleared, would be ſufficient to indemnify him for the loſs in quantity. Thoſe who mean to try this experiment would do well to advert to the following particulars:

1ſt. To the equality of the ſize of the ſeeds at planting.

2dly. To

2*dly*. To the progress of the growth of the plant at each period.

3*dly*. To the different kinds of potatoes with which it is tried. And,

4*thly*. To the nature of the weather at the time.

Experiment Seventh.

With a view to ascertain whether any loss, and what, as to the weight of potatoes, was sustained by cutting over the stems at different periods, all the plants whose stems were cut over in experiment 5th, were allowed to stand till the 28th of October, at which time they were all taken up, and the produce of each parcel separately weighed. On the 28th of October also, nine other plants marked x x x, &c. see the foregoing diagram, being part of three rows, three plants in each row, that grew contiguous to the potatoes cut, (two rows intervening, so as that they could not be influenced by the opening occasioned by cutting the stems of the potatoes o o o, &c.) were taken up by themselves and separately weighed. This, it was supposed, gave a very fair average of what the *cut* plants would have yielded, had they been allowed to remain uncut; and of course, that the difference between the weight of each of these patches shewed the loss of crop occasioned by the cutting of the stems at the different periods indicated.

These particulars are expressed in the following table;—to which is added a column, denoting the total loss of crop *per acre*, that would be sustained by cutting over the stems at each period respectively.

Time

Time when the ſtems were cut over.	Produce of nine plants *cut over*; taken up October 28th.		Produce of nine plants *uncut*; taken up Oct. 28th.		Difference between the produce of nine ſtems *cut* and *uncut*.	Loſs of crop *per acre*, occaſioned by cutting over the ſtems.	
	Weight. *lb. oz.*	Num.	Weight. *lb. oz.*	Num.	Weight. *lb. oz.*	Wt. *lb. dec.*	Buſhels.
Auguſt 2d	2 12	78	12 12	101	10 0	35.000	624
Auguſt 10th	5 8	100	13 11	96	8 3	28.650	511
Auguſt 17th	6 2	90	13 12	94	7 10	26.691	476
Auguſt 22d	9 5	103	13 13	97	4 8	15.750	281
Auguſt 29th	10 10	110	14 1	100	3 7	12.031	214
September 5th	12 0	102	13 8	96	1 8	5.250	93

From this experiment it appears, that if the ſtems of this kind of potatoe be cut over about the time they are coming into bloſſom, there would be a diminution of the crop of ten parts out of twelve nearly, of the whole produce, or a loſs at the rate of 624 buſhels *per acre*; and that a proportional loſs would be ſuſtained by cutting the ſtems at any future period of their growth. Hence it is obvious, that the loſs by this practice would be much greater than could be counterbalanced by any advantage that could be derived from the green ſtems, as feeding for domeſtic animals.

Though it alſo appears from this experiment, that potatoes advance but very little after the ſtems are cut over; yet, by comparing this experiment with the former, it would ſeem that they did advance a little. This may be inferred from the following table;—the firſt column of which is the reſult of experiment 6th, multiplied by nine—and the laſt is taken from experiment 7th.

	Produce of nine plants taken up at the periods mentioned in the margin.	Produce of nine plants cut over at the ſame periods.	Difference; being the increaſe after the ſtems were cut.
	lb. oz.	*lb. oz.*	*lb. oz.*
Auguſt 2d	2 1¼	2 12	0 10¾
Auguſt 10th	3 15	5 8	1 11
Auguſt 17th	5 5¼	6 2	0 12¾
Auguſt 22d	9 4¼	9 5	0 0¾
Auguſt 29th	7 5	10 10	3 5
September 5th	12 15	13 8	0 9

Though

Though I muſt again obſerve, that no accurate concluſion can be drawn from the reſult of experiment 6th; the unuſual great produce of the ſtem taken up Auguſt 22d, and the ſmaller produce of that of the 29th, were probably owing to the larger ſize of the cutting in the firſt than in the laſt, or to ſome other unobſerved circumſtance. This compariſon therefore only affords a probable reaſon to ſuſpect, that the plants do increaſe ſomewhat after they are cut over, though but a little.

Experiment Eighth.

To aſcertain whether a benefit might in any caſe accrue from replanting the ſtems of potatoes that were taken up for an early crop, and to what that might amount, the ſtems of the different plants that were taken up in experiment 6th, were all immediately replanted after the bulbs were taken off and weighed. It was found that theſe plants readily took root, and produced another crop of potatoes that ſeaſon, the amount of which, when ſuffered to remain in the ground till the 28th of October, and the proportion that this ſecond crop bore to the firſt, is denoted below: one ſtem being in both caſes multiplied by nine, to admit of their being the more readily compared with the reſult of the other experiments mentioned in this ſection.

	First produce from 9 ſtems, taken up at the times denoted on the left hand.	Second produce of nine ſtems replanted.	Difference between the firſt and ſecond produce.	Total produce of both plantings from nine ſtems.
	lb. oz.	*lb. oz.*	*lb. oz.*	*lb. oz.*
Auguſt 2d	2 1¾	*Torn up by accid.*		
Auguſt 10th	3 15	2 11½	1 3½	6 10½
Auguſt 17th	5 5½	1 2	4 3½	6 9
Auguſt 22d	9 4½	0 9	8 11½	9 13½
Auguſt 29th	7 5	0 4½	7 0½	7 9½
September 5th	12 15	0 4½	11 10½	13 3½

It thus appears that a ſmall quantity of potatoes may be obtained by replanting the ſtems, if taken up at a very early period;

period; yet this, at the beft, is but a trifling acquifition, and probably can never in any cafe be worth the expence; efpecially when it is alfo adverted to, that the fecond produce of potatoes thus gained are always bulbs of fo fmall a fize, as to be of very little value in proportion to their weight. It may be a fatisfaction, however, to fome to know, that in cafe a ftem of a particular kind, of which one has very few, be pulled up by accident before the potatoes are fit for feed, it need not be altogether loft, but that by replanting it the kind may be ftill preferved.

Where a few early potatoes are wanted, the moft œconomical practice is to pick out with the fingers (which may be eafily done in a well-dreffed foil) thofe bulbs that have attained the fize fit for ufe, leaving the ftems in their place, with the fmall potatoes upon them to grow till they alfo attain a fize fit for being ufed.

§. II.

The foregoing experiments were all made with one kind of potatoes; nor have I taken notice above of any other varieties. But as there is a very great diverfity in this refpect, and as the properties of one kind are often very different from thofe of another, our knowledge of the value, and moft proper mode of cultivating this plant, muft be very incompleat, until the farmer fhall know the different weight of crop, &c. that he could obtain by cultivating any one kind in preference to any other; for different forts are known to vary very much from each other in regard to prolificacy, as well as in

feveral

ſeveral other reſpects. A neceſſary ſet of experiments therefore would be a comparative trial, in equal circumſtances, of all the different kinds, with a view to aſcertain the weight of produce that could thus be obtained from each.

The only experiment under this head I ever made, was the following:

Experiment Ninth.

At the ſame time that the potatoes in experiment 2d were planted, I made choice of twenty plants of a different kind of potatoe, that is uſually diſtinguiſhed in Aberdeenſhire by the name of the *Dutch cluſter potatoe*. The bulbs of this kind are of an irregular roundiſh knobby form. The colour of the ſkin a yellowiſh white, with a faint pinkiſh tinge about the eyes, eſpecially before it is quite ripe. The colour of the pulp yellowiſh white—conſiſtence viſcid, not meally: taſte ſweetiſh. The eyes are pretty deeply ſunk in the bulb. The ſtalk and leaves are neither ſo long, nor ſo dark in the colour, as moſt other kinds; but are generally numerous, more erect, and leſs jointed. Bloſſoms white, with a pale pink or purpliſh tinge, numerous and large. Theſe are ſucceeded by apples, which in this kind are generally abundant, and of a large ſize. The umbilical fibres do not in this kind ſpread to any conſiderable diſtance from the ſtem, ſo that the bulbs are uſually found in a cluſter cloſe to the root, to which they firmly adhere. It is reckoned a great bearer.—Theſe are the principal characteriſtics that at preſent occur to me for diſtinguiſhing this kind, which I only do from memory.

Twenty whole potatoes of this kind, which weighed exactly 123 ounces, were planted in a row immediately contiguous

contiguous to the eighth row in experiment 2d, (which weighed alſo 123 ounces) at the ſame time with them, and both rows were managed in every reſpect exactly alike. They were alſo taken up, and the produce weighed at the ſame time, when the weight of each was found to be as under:

	lb.	oz.
The 8th row in experiment 2d, conſiſting of the white kidney potatoe—the produce weighed - - - - -	21	5½
The row of Dutch cluſter potatoes - - -	27	1
Difference - -	6	11½

Which is equal to about 150 buſhels per acre in favour of the Dutch cluſter potatoe. Though no abſolute dependance can be had on one experiment only, yet it plainly appears, that much benefit might be derived from the experiments propoſed in this ſection, if properly executed.

§. III.

The reader will pleaſe to take notice, that all the experiments above recorded (thoſe in ſection 1ſt, paper 2d, only excepted) were made upon a poor undunged ſoil, for the ſake of accuracy; ſo that the crop, upon the whole, was very poor. I have never yet had a proper opportunity of making any trials that could with accuracy aſcertain what might be the greateſt crop that could be obtained from an acre: nor indeed can that point be fully aſcertained, till the experiments ſuggeſted in ſection Xth,

Xth, as well as to comparative trials mentioned in the laſt ſection, with ſome others, ſhall have been made. From ſome trials I have made, but with leſs accuracy than to admit of being here recorded, I have reaſon to be ſatisfied that the *poſſible* produce from an acre is much greater than moſt perſons at preſent imagine to be obtainable. I mean to proſecute theſe experiments next ſeaſon, if I am not prevented by ſome unforeſeen accident, and ſhall not fail to communicate the reſult to the publick in due time. But though my intention is to try to elucidate this ſubject myſelf, I beg leave warmly to recommend it to others alſo; for it is impoſſible that a matter of ſo much importance can be too fully inveſtigated.

§. IV.

The reader who has attended to the accounts that have been publiſhed of the various crops of potatoes that have been obtained by different perſons in different ſituations and circumſtances, cannot fail to have obſerved, that the diverſity in the total produce *per acre*, is much greater than can well be accounted for, by any particulars of the ſoil or culture that have been taken notice of: ſuch a diverſity, however, will now no longer appear wonderful, when he remarks, that the *ſize* of the ſeeds planted has never in any inſtance been ſufficiently

adverted

adverted to: ſo little indeed has this been done, that is is only caſually that it is taken notice of at all; though the foregoing experiments clearly prove it to be of the moſt eſſential importance with reſpect to the total amount of the crop.

PAPER THIRD.

OF ARDENT SPIRITS AFFORDED BY POTATOES.

§. I.

THE uſes of the potatoe as a food for man, and the domeſtick animals he rears, are already pretty well known; but it is not in general underſtood that from this plant may alſo be obtained a vinous ſpirit, of an excellent quality, in very large proportions. A good many years ago an account of an experiment made in Sweden to aſcertain this fact was publiſhed in the memoirs of the Philoſophical Society of Stockholm. This, together with ſome obſcure hints I received from the late ingenious Dr. John Gregory, of ſome ſimilar experiments that had been made with ſucceſs in the North of Scotland, induced me to make the following trial.

EXPERIMENT

Experiment Tenth.

February 15th, 1777, I set apart two *Aberdeenshire* pecks of potatoes by measure, which I have since found were each equal to 36 pounds by weight, so that the whole was 72 pounds. These potatoes were boiled in a cauldron till they were brought to a soft pulpy state; they were then bruised, and made to pass through a strait riddle along with some fresh water; the skins being kept back by the riddle, which were thrown away. The pulp was then mixed with cold water, till the whole amounted to about twenty gallons English. This was allowed to cool till it attained the same temperature as would be proper for mixing yeast with wort; when some yeast was put to it, as if it had been yeast to wort from malt. In ten or twelve hours a fermentation began, which continued very briskly for the space of ten or twelve hours, but at the end of that time it began sensibly to abate; from which circumstance I was afraid my experiment would fail. After waiting for some time, and in vain, warming it a little, with a view to renew the fermention, I determined to stir it briskly to see if it could be renewed by that means. This produced the desired effect, and the same operation was renewed every day, and the fermentation continued to go on in a proper manner for a fortnight. At the end of this time the fermentation abated, and could not be renewed by agitation or otherwise; and the liquor, having been found upon trial to have acquired a kind of acid, slightly vinous taste, was judged fit for distillation. It was then distilled with due caution, care having been taken to stir it in the still, until it began to boil, before the head of the still was applied; and the fire was afterwards kept up so strong as to keep it boiling briskly till the whole was run over. This was intended to prevent the thick matter from subsiding to the bottom; for I was afraid that without this precau-

tion,

tion, it would have acquired a *still-burnt* flavour; and I found by experience in one instance, that this kind of *empyreuma* was of an exceeding disagreeable kind, resembling in flavour the fumes of burning tobacco.

In consequence of these precautions and due rectification, I obtained an English gallon of a pure spirit, considerably above proof; and about a quart more of a weaker kind, a good deal below proof. This was, in every respect, the finest and most agreeable vinous spirit I ever saw. In taste it somewhat resembled very fine brandy; but it was more mild than any brandy I ever tasted, and had a certain kind of coolness upon the palate peculiar to itself, by which it might be readily distinguished, by a nice judge, from every other kind of spirit. Its flavour was still more peculiar to itself, but it more nearly resembled brandy impregnated with the odour of violets and rasberries, than any thing else to which I could compare it. [A single glass of it put into a bowl of rum punch, made it appear as if it had consisted half and half of rum and brandy, impregnated with the juice of raspberries.] It seemed to derive this flavour from a subtile essential oil, of a very singular kind—for although it rose with the first spirit that came over, it still continued to come over, without any sensible diminution or change of flavour, till the whole of the spirit was entirely drawn off. It was also so difficult to be dissipated, as to scent with its own perfume a drinking glass, into which the spirit had been poured, for more than twenty-four hours after it had been emptied, and apparently quite dry; and this perfume, after the spirituous flavour was totally dissipated, appeared to me the most agreeable I had ever met with. I have been at the greater pains to describe this kind of spirit in its state of perfection, because I have since heard of and seen some spirits, said to be drawn from potatoes, which, for want of ſkill or caution in the operators, was intolerably nauſeous. As others may fall into the ſame errors in attempting to perform the ſame operation, I ſhall hazard a few remarks on the cautions neceſſary to be obſerved in attempting to extract vinous ſpirits from this or other roots; for want of attending to which particulars, many attempts of this kind have no doubt failed.

§. II.

Every philoſophic enquirer knows that vinous ſpirits are entirely the produce of fermentation, and cannot be obtained from any ſubſtance whatever, till it has undergone that chemical proceſs: but many of thoſe who attempt experiments of this kind, are neither ſufficiently aware of the neceſſity of this previous ſtep, nor acquainted with the means of exciting it, or of conducting it properly, which frequently fruſtrates their attempts.

If any vegetable in an unfermented ſtate be diſtilled, there is, for the moſt part, obtained by that operation, a portion of native *eſſential oil*, ſtrongly impregnated with the peculiar taſte or flavour of the ſubſtance from which it is obtained: but if the ſubſtance be properly fermented, that eſſential oil diſappears, and in its ſtead a new ſubſtance is obtained by diſtillation, altogether different from the former in many reſpects. This ſubſtance is called *vinous ſpirits*, or *alcohol*, when in its higheſt rectified ſtate.

But

But if any vegetable substance be subjected to distillation before it has been made to undergo a *proper degree* of fermentation, a *part* of it only rises in the state of *vinous spirit*, and a *part* of it also rises in the state of *native essential oil*; which, mixing with the spirit while in the state of vapour, and being dissolved therein, communicates to that spirit a taste and flavour very different from that of the pure spirit by itself, which is, for the most part, extremely nauseous and disagreeable. It has pretty much the same effect as if a quantity of the raw vegetable substance should be distilled along with another quantity of it that had been *properly* fermented. In all those cases, where the volatility of the native essential oil is nearly the same with that of the spirit, it is evident that no care in the process of distillation can prevent them from being blended together in the same process.

From hence it appears sufficiently obvious, that if ever we hope to obtain the pure genuine vinous spirit without adulteration from any vegetable substance whatever, it is of the very greatest consequence that the fermentation be properly carried on, so as that the whole of the matter susceptible of fermentation shall be equally and entirely assimilated before it be committed to the still. This is on all occaſions neceſſary; but it is peculiarly ſo in thoſe caſes in which the native oils are very abundant, or volatile, or diſagreeable. In diſtilling malt ſpirits, this circumſtance is ſeldom ſufficiently attended to; the fermentation being uſually hurried forward with a rapid careleſſneſs, in conſequence of which ſome part of it is converted into vinegar, before other parts of it are aſſimilated at all. Hence it neceſſarily follows, that the malt not only yields a ſmaller *quantity* of ſpirit, but affords that ſpirit alſo of a much inferior *quality* to what it would have been if the fermentation had been duly conducted. Spirits that are drawn from ale, which has been accidentally allowed to run into the acetous fermentation, are always, on this account, of a quality far ſuperior to that obtained from malt by any other proceſs.

In attempting therefore to obtain a ſpirit from roots or other vegetable ſubſtances, the firſt point to be attended to is, to conduct the fermentation properly, and to puſh the vinous fermentation as far as it can be made to go. I am diſpoſed to aſcribe the ſucceſs I had in this experiment, beyond what others have experienced, in a great meaſure to this cauſe, and to the care that was taken to prevent it from obtaining the ſlighteſt empyreumatic taint during the diſtillation; though it may alſo have been occaſioned by ſome other unobſerved peculiarity.

One

One particular I remarked relating to the diſtillation of this ſpirit, that deſerves to be mentioned. In diſtilling from malt, it is found that towards the end of the operation a quantity of weak ſpirit is forced over, which is ſtrongly impregnated with a very diſagreeable oil, that very much debaſes the whole of the ſpirits, if it be ſuffered to mix with them. To ſeparate this from them, with as little loſs of good ſpirit as poſſible, conſtitutes one of the principal niceties in the proceſs of diſtillation from malt. But no ſuch phenomenon occurs in the diſtillation from potatoes; for I could perceive no difference between the taſte of the very weakeſt ſpirit towards the end of the operation, and that which came over at the beginning or any other part of the proceſs, if equally diluted with water. It would ſeem that the oil, to which this ſpirit owes its fragrance, is in all parts of the proceſs ſeemingly the ſame, and always agreeable; contrary to the *gout*, or *goo*, as it is pronounced, of malt.

§. III.

I have deſcribed above, with all the accuracy I could, the whole proceſs and phenomena that occurred in diſtilling ſpirits from potatoes, as I obſerved them when the proceſs was conducted under my own eyes. This proceſs I repeated twice, about the ſame period of time, with the ſame ſucceſs. But it is alſo juſt to obſerve, that though it has been ſince that time ſeveral times attempted by my direction, under the care of another perſon, on whoſe accuracy I thought I could depend, it has invariably failed in as far as reſpects the peculiar fragrance of the ſpirits above deſcribed, though in every other reſpect the reſult was the ſame with mine: the ſame yield of ſpirit of equal ſtrength being obtained, which was diſtinguiſhed by the ſame cool ſenſation on the palate, and in every reſpect an excellent ſpirit, though diveſted of that unuſual fragrance above deſcribed. I have often wiſhed to repeat the experiment myſelf, and ſo to vary circumſtances as to try to diſcover the cauſe of this peculiarity; but the revenue laws are ſo ſtrict at preſent, that a private man cannot venture to have a ſtill in his poſſeſſion for the ſake of making any experiment of this ſort, without ſubjecting himſelf to a very heavy penalty; and as I ſhould very much diſlike any thing that had the appearance of evading the laws, I have thus been, very much againſt my will, prevented from repeating theſe experiments. Certain, however, as I am with regard to the fact, (which if neceſſary could be atteſted by many perſons who taſted the ſpirits) I have no ſcruple in publiſhing it fairly to the world, leaving it to time,

and

and to others who have opportunity to make these experiments, to discover the causes of this peculiarity, and other particulars relating to it.

If the vegetable substance that is subjected to fermentation contain but a small proportion of fermentable matter, it will not be possible ever to free the spirits from the peculiar flavour of the vegetable; for that large proportion of unassimilated matter being subjected to distillation, along with the fermented liquor, will of necessity yield its oil by the heat employed to distil the spirits. This seems to be particularly the case with regard to carrots, parsnips, and turnips, all of which I have tried, and found that although they could be made to undergo the process of fermentation, and to yield a considerable proportion of ardent spirits, yet that these spirits were strongly tainted with the flavour of the vegetables from whence they were obtained, and so intolerably nauseous, that they never could be employed for food by man. In the process above described, the whole of the matter of the potatoes was subjected to distillation. What effect would have been produced by separating the gross sediment from the transparent fluid above it, after the fermentation was over, either as to the quantity or quality of the spirit, I had not an opportunity of remarking; but ſhould ever the proceſs of extracting ſpirit from potatoes be attempted on a large ſcale, it would be of importance to try to ſeparate that ſediment before diſtillation, as that proceſs would be rendered much eaſier, and leſs precarious, in conſequence of that operation.

If ever this manufacture ſhould be attempted, it deſerves alſo to be remarked, that the farinaceous powder which ſubſides to the bottom after the fermentation, ſeems to have ſuffered very little change in its taſte or appearance by the proceſs, as it very much reſembles boiled potatoes in all reſpects, ſo that it might probably go as far, as food for domeſtic animals, as the potatoes themſelves would have gone in their native ſtate.

I ſhall only farther add on this ſubject, that I attempted to obtain a fermentable liquor, by bruiſing the potatoes raw, and pouring water of different degrees of warmth upon it, as is uſed in maſhing malt, but could never thus ſucceed in exciting any degree of fermentation. It always afforded a viſcid roapy liquor, that remained unaltered after the addition of yeaſt to it. I now return from this long and intereſting digreſſion.

PAPER FOURTH.

OF THE MARKS FOR DISTINGUISHING DIFFERENT SORTS OF POTATOES FROM EACH OTHER:—ITS UTILITY, &c.

§. I.

I Have had occaſion to obſerve, in ſome of the foregoing parts of this eſſay, that there are ſeveral varieties of potatoes, which differ from one another conſiderably in ſome of their moſt eſſential properties. Theſe varieties are indeed ſo numerous as renders it impoſſible for almoſt any perſon not to have remarked them, yet no one is ſo well acquainted with all theſe varieties and their properties, as to know with certainty which kind would be moſt profitable to cultivate on every particular occaſion; for want of which knowledge, much loſs muſt be annually ſuſtained by the public. But till ſome method ſhall be adopted for diſtinguiſhing each kind from another with certainty, it is in vain to hope for any comparative trials that could be of material utility to the farmer. To begin this ſyſtem of claſſification as to this particular, the following hints may be of uſe.

As it is impoſſible to convey a diſtinct idea of the ſmall variations that require to be here attended

to,

to, in so easy a manner as by comparing every variety with one kind that shall be considered as an universal standard, to which all descriptions should refer; the first step therefore will be to fix on one kind that shall be proper to be considered as an universal standard. The difficulty is to find a kind that may be distinguished from all others, by such striking characteristicks as to prevent a possibility of mistaking it.

In casting about with this view, it seems to me that the kind known in Scotland by the name of the *wise* potatoe, promises to answer this purpose better than any other, because it is distinguished from other kinds by *one* very obvious peculiarity, viz. that of never carrying any blossom or fruit. [Since the above was written, I have seen some other sorts that carry no blossom, but these may be very easily distinguished from it by some other of its obvious characteristicks.]——Its peculiarities are as under. In form the bulbs of this kind are remarkably regular, being all of the shape of a heart, somewhat longer than its due proportion, and flatted a little one way. The fibre by which it adheres to the stalk, which I would call the *umbilical* cord, adheres to the great end of the bulb, and the point of the small, and is thickly covered

with

with eyes. The skin is smooth and thin; its colour a pale red, rather brighter at the point than elsewhere. Its flesh is of the meally sort; the taste rather tending a little to sweetishness. The fibres, to which the bulbs adhere, do not ramble very wide, nor do they keep so close to the stem as some other kinds; they neither push very deep, nor rise extremely near the surface. The bulbs themselves are never remarkably small, nor uncommonly large, but of a good equal size, and it is reckoned a good bearer.

By attending carefully to these marks, it might in general be well known; and when any person was once possessed of a plant or two of this standard kind, with which all others could be compared, he would thus be able to point out with accuracy the smallest discriminating peculiarity, so as to be in no danger of mistaking any others that should be the object of discussion.

This fundamental step being first taken, I would recommend, that in every description attention should be given to specify all the following peculiarities :—

Below Ground.

The general form and size of the bulbs.

Their colour.

The

The ſmoothneſs or roughneſs of the ſkin.

The conſiſtence, that is, the mealineſs or viſcoſity, and taſte, of the bulb.

The colour, length, thickneſs, &c. of the umbilical cord.

Their tendency to go deep, or to riſe near the ſurface; to ramble wide, or to adhere cloſe to the ſtem.

The time when the bulbs knot and ſet; marking, not by the kalendar only, but alſo compared with the advance of the plant above ground.

The time when they attain perfect maturity with reſpect to ſize, and alſo that period of their growth at which they loſe the herbaceous, and attain the farinaceous taſte.

Their general prolificacy.

How long they may be kept, at what ſeaſon they are in greateſt perfection for eating, &c.

Particulars obſervable above ground.

The general height, colour, and form of the ſtem.

Their tendency to puſh out many or few ſtems from a root.

Whether they carry bloſſom or not.

The form, dimenſions, and colour of the leaves.

The form, colour, and general habitude of the bloſſom, where there is any.

The time at which the bloſſom appears.

The tendency they have to produce few or many apples.

The tendency they have to produce thoſe excreſcences on the ſtalks that reſemble potatoes below ground, which may be called air potatoes.

The

The comparative hardineſs or tenderneſs of the leaves, in reſpect of froſt or other variations of weather that affect them.

Particulars that concern the whole plant.

The ſoil which ſeems beſt to ſuit each kind.

The mode of culture that beſt agrees with them.

The accidents which are moſt liable to affect them; and in general every particular that could indicate any difference between one kind and another.

§. II.

Thoſe who have not been accuſtomed to attend to the growth of this plant, will perhaps think that ſome of the above marks are of no moment: and ſome of the particulars they will not be able to underſtand. To obviate theſe objections, a few explanations are neceſſary.

The potatoe, becauſe it grows below ground, has been uſually called a *root*—but improperly. It more nearly reſembles a kind of underground fruit; and in conformity with this idea, the French have given it the name of *Pomme de Terre*, ground apple. This fruit has a ſet of organs peculiarly adapted for its production, in the ſame manner as every other kind of fruit above ground has a ſet of organs for their production; which organs appear at the

proper

proper period, carry the fruit, and then decline, in a manner exactly analogous to what happens below ground with the potatoe. The potatoe plant, when it begins to vegetate, sends forth roots into the ground, by which it imbibes its nourishment like every other plant; but after it has arrived at a certain period of its growth, it begins to shew its fruit, bearing apparatus below as other plants do above ground. This below ground consists of a set of fibres quite distinct from the roots, which are at first of a tender fleshy consistence, usually of a whitish colour, which is in some kind blended with a slight tinge of red. These gradually extend themselves around the plant to a greater or smaller distance in different kinds of potatoes, and from these in due time spring out the bulbs, or fruit, appearing at first like small excrescences upon the fibres, which gradually expand, and assume their proper shape as they advance towards maturity, very much resembling, in these particulars of their growth, the progress of the cones of the larix tree upon its small fibrous branches. These fruit-bearing fibres become by degrees less bright in colour, and more firm in consistence; and assume a dark colour and stringy consistence, as they advance towards perfection. This set of fibres I would distinguish by the name *umbilical*, from the great similarity in

office

office they bear to the animal organ so called; and because they never yet have obtained an appropriated name.

Different kinds of potatoes do not differ from one another more in any one respect than they do in the form, colour, habitude, time of springing forth, &c. &c. of this apparatus of fibres; so that this ought not only to be attended to as one mark of distinction between different kinds, but also as a particular that may on some occasions influence the mode of culture that would be proper for particular kinds. I shall give one example.

It is found by experience, that some kinds of potatoes may be profitably cultivated by means of the horsehoing husbandry; (possibly under due regulations this might be always of use) but in some cases that mode of culture is attended with danger; for, should the kind of potatoe that is thus cultivated have a tendency to send out these umbilical fibres early and to a great distance, if the plough should be employed after these were shot forth, it might cut them off, which would have a very different effect from cutting the roots that absorb food for the plant. The stems might thus indeed be increased, but the produce in fruit much diminished. I have seen a field of horsehoed potatoes, which, owing to this circumſtance, although a very luxuriant crop above ground, yielded when taken up only a very few well-formed bulbs; the umbilical fibres being at that time in a ſucculent growing ſtate, and covered with ſmall crude potatoes that would have required a very long time to bring them to maturity. Late and *deep* hoeing, even with the hand-hoe, is, on this account, with ſome kinds of potatoes highly pernicious.

On the other hand, there are ſome kinds of potatoes that never ſend theſe umbilical fibres above a few inches from the ſtems, which would not be liable to the ſame objection, as there would be no danger of having them cut by the plough; and other ſorts ſend theſe fibres directly downward to a great depth, ſo as to be in no danger of being in any caſe wounded by the hand-hoe.

Other kinds of potatoes have a tendency to ſend out bulbs at every joint of the ſtem, even above ground; but unleſs theſe be covered with earth they never acquire the colour or taſte of real potatoes, although they have the exact ſhape and appearance. I have ſeen ſome ſtems of potatoes, eſpecially in a rainy ſeaſon, that were covered with theſe green potatoes to the very top, and have numbered fifteen or ſixteen on one ſtem, ſome of them

of

of the ſize of ſmall hen's egg's: [And I once met with a cluſter of that kind of potatoes, conſiſting of about twenty bulbs, that occupied the place of potatoe-apples, ſpringing all from one foot-ſtalk that adhered to the ſtem, preciſely in the ſame way with that which ſupports the bloſſom and ſeed veſſels.——This I have ſtill in my poſſeſſion.] Nature ſeems here to indicate, that the ſtems ought to be covered in part with earth to blanch them, (potatoes that grow below ground, if laid bare while in their growing ſtate, aſſume the ſame green appearance) which would probably in theſe kinds augment the crop conſiderably; although with regard to ſuch kinds as have no tendency to produce bulbs along the ſtem, the operation of covering them would probably be much leſs beneficial. Care, therefore, ſhould be taken to ſelect the firſt of theſe kinds of potatoes, where it is intended to rear them, after the Iriſh faſhion, in *lazy beds*.

I have mentioned this tendency to produce potatoes on the ſtems above ground, as one means for diſtinguiſhing different kinds from each other; for although a few kinds are endowed with this quality of producing bulbs above ground, in the ſame way as ſome peaſe that produce pods both above and under ground; yet this in the one caſe as well as

 the

the other ſeems to be contrary to the ordinary œconomy of both kinds of plants, and therefore ſerves as a proper mark of diſtinction.

§. III.

The potatoe admits of being tranſplanted as eaſily as moſt other plants, eſpecially if this be done before the umbilical fibres ſpring out. Doubtleſs this property might be laid hold of with advantage for cultivating thoſe, eſpecially of the early ſort; though I do not know that it has ever yet been attempted to be carried into practice.

PAPER FIFTH.

OF RAISING POTATOES FROM SEED.

§. I.

IT is not many years ſince it was firſt diſcovered that potatoes could be reared in Europe from actual ſeeds, the produce of our own climate; but this fact is now aſcertained without the poſſibility of a doubt. As many improvements have been ſaid to reſult from this mode of culture; and as the deſcriptions hitherto given of the effects that reſult from

from it are lame, and have been delivered with such a mysterious air, as to give me no distinct notion of the matter; I resolved to satisfy myself experimentally on that head, which was done as under:

Experiment Eleventh.

Upon the 23d day of April 1776, I sowed, on a bed of good garden mould, some seed potatoes that had been gathered the former autumn, and had been preserved among some dry straw all the winter, to prevent them from being injured by the frost. The apples, which had been packed up whole, were by that means so much dried, that I found it a difficult matter to separate the seeds sufficiently, which occasioned the plants to come up in tufts much thicker in some places than others. The young plants appeared above ground in about ten days, and advanced vigorously during the summer, especially in those places where they were not too thick. On the 3d day of November thereafter, they were carefully taken up, when it was found that some of them were nearly as big as a pigeon's egg, gradually decreasing from that to the size of common pease, many of them being no larger. A few of the largest of these were boiled, and others roasted, with a view to discover if they possessed that rich almond-like taste, which some persons had said the potatoes raised from seeds always possessed in a remarkable degree. They were found to eat very well, but not one bit better than other good potatoes of the same kind that had been raised from sets in the usual way. The remainder were carefully packed up to guard against frost, and were thus preserved for planting in the spring.

April 20th, 1777, these small potatoes were planted in a bed of good garden mould, in rows one foot asunder, and at four inches on an average apart in the rows. On this occasion I began to plant at one end of the bed, the rows going across it, and proceeded regularly towards the other end, always selecting the largest bulbs I could observe. By this means it naturally happened, that the biggest plants were all placed at one end, and gradually diminished towards the other end of the bed, where the very smallest were planted; and as the last of these came to be very small indeed, I gave them less room in the rows, decreasing, as the plants diminished in size, from six to two inches. All these plants were equally cared for during the summer; but it was observable that the stems which grew from the largest plants were from the beginning exceedingly large, luxuriant, and healthy, in comparison of the smaller ones. The leaves of these were broad and healthy, and the whole plant above ground appeared at least *ten times* greater than the puny plants that sprang from the small seeds. They were all taken up in the month of October, when it was found that the largest seeds yielded a good crop of potatoes, many of which were as big as a hen's egg; but those produced from the smaller seeds did not in general exceed the size of a horse-bean, and many of them no bigger than small pease. None of the plants shewed blossom this season. The bulbs were carefully preserved for planting in the spring.

[These potatoes were accordingly planted in the month of April 1778, in rows about a foot from each other, and the largest were planted at the distance of one foot in the rows, the smaller being placed closer as their size diminished, so as that the least stood about four inches apart in the row. The largest seeds again produced by far the most luxuriant

luxuriant and weighty crop, and a few, and but a few of them, ſhewed any bloſſom; but none of the bulbs, not even thoſe from the largeſt plants, were nearly of ſuch a ſize as thoſe produced from very large potatoes: nor did they afford nearly the ſame produce per acre as was obtained from old potatoes planted on the ſame ſoil at eighteen inches apart. The facts wiſhed to be eſtabliſhed by this experiment being now aſcertained, and I being engaged in other intereſting purſuits, it was not thought neceſſary to continue it longer. The following corollaries ſeem to be clearly deducible from it.]

§. II.

From the accounts I had received of potatoes raiſed from ſeed, it did not appear to me clear whether new bulbs were produced from theſe potatoes in the ſecond or third year of their growth, or whether theſe potatoes during that time continued only to increaſe in bulk, without producing other potatoes from them. It was always ſaid that they did not *attain perfection* till the third year from the ſeed; and what was meant by their attaining *perfection*, I could not gather from any accounts I had ſeen. And as it was ſaid they could be obtained by this means much earlier in the ſeaſon than others, and were poſſeſſed of many other ſingular qualities, I could not tell what judgment to form of it. From the foregoing experiment, however, it clearly appears, that after the firſt year theſe ſeedling potatoes puſh forth ſtems and bulbs exactly in

 the

the ſame manner as any other potatoes planted for ſeed, and agree entirely with them in other reſpects; the largeſt in this caſe, as in every other caſe, as in every other caſe, whether cuttings of old potatoes or whole ones be planted, always producing the largeſt bulbs and the moſt weighty crop.

As to the notion of their attaining their full ſize on the third year, and not before, this ſeems to have originated merely from inaccurate obſervation. It does not ſeem poſſible to aſſign any preciſe period at which theſe bulbs will invariably attain perfection, as that muſt in general depend on many accidental circumſtances. It appears that the ſize of the bulbs produced in the ſecond and third year depend in a great meaſure on the bigneſs of thoſe that were planted; and that this will be influenced by the richneſs of the ſoil, and the diſtance allowed to the ſeedling plants the firſt year. I know no circumſtance that could ſo well be aſſumed as ſuch a probable criterion of the potatoe having attained perfection, as that of its puſhing forth flowers and producing ſeeds *properly ſo called.* Now, although none of the plants in my experiment produced flowers in the ſecond year, yet it is not improbable, that on ſome occaſions, if the ſeeds were ſown very thin and on a rich ſoil, the bulbs of the firſt year's growth might be much larger than any of thoſe

raiſed

raised by me;—so neither is it in the least improbable, that in that case some of the best of them might produce blossom in the second year. On the other hand, as the bulbs of the second and even of the third year's produce, produced from the smallest plants, were some of them not so large as some of those of the first year's growth, and as the vigour of the plant, and the size of the bulbs, and quantity of blossom produced, evidently depend on the size of the potatoes planted, it is probable that these small bulbs would require a year longer than the former to attain the same symptoms of maturity. In short, as the vigour of the future plant, &c. seems in this case to depend very much upon the size of the bulbs planted for seed, it is probable that if two plants of very unequal magnitude were picked off from the same stem, and planted out as seeds, the one of them might be found to have attained its full degree of perfection, so as to carry blossoms and fruit in abundance, while the other yielded none at all; and if the same process were repeated, the same phenomena might be produced *in infinitum*. The *age* therefore of the plants, by which we must here be understood to mean the number of years from the time that the seeds were sown, can give no precise indication of the state of the crop that may be expected from them, independent of the size of the bulbs.

Although,

Although, in compliance with the uſe that others have made of the term, I have ſpoken of potatoes attaining a period of perfection that has been denominated *maturity*, I muſt here enter a caveat about this application of the term, as being indefinite and inaccurate. I ſaw no room to ſuſpect that the potatoes raiſed from ſeed had not in the firſt year, though ſmall in ſize, attained as great maturity; that is, in proper circumſtances, were as well ripened, and as fit for uſe, as others of the ſame ſize ever afterwards would become. Nor are the largeſt potatoes of the ſame kind, if taken from the ſtem at the ſame period of its growth, ſeemingly much if any thing ſuperior in taſte, and other qualities that could indicate *maturity*, to thoſe of a ſmaller ſize. The term *maturity*, therefore, in the ſenſe above given, ſeems to be extremely improper, and ſhould be entirely confined to denote the degree of ripeneſs that the plants had attained in the ſeaſon they are produced in. It ſeems that a certain *ſize* of potatoes planted for ſeed, the ſoil and climate given, is neceſſary for producing a plant of ſuch vigour as will puſh out bloſſoms and fruit, and that till it attains that degree of vigour *at leaſt*, it never produces bulbs below ground of the largeſt ſize. But what the other circumſtances are which tend to augment the ſize of the bulb to the greateſt poſſible degree, we cannot as yet poſſibly ſay.

§. III.

In conformity with the notion that *precisely* three years are neceſſary for bringing to perfection the firſt bulb raiſed from ſeed, it has been aſſerted, that no method is ſo proper for obtaining potatoes very early as to plant theſe ſeedling potatoes after the ſecond year's growth pretty early in the third ſpring, to ſuffer them to remain in the ground for ſome time, and to dig them up in the months of June or July; when, it is ſaid, they have attained their full ſize and due perfection in every reſpect. But I did not find from experience the ſmalleſt foundation for this beautiful hypotheſis. It has already been ſaid, that no augmentation in the ſize of the parent bulb takes place, after being planted, in this caſe more than in others. The parent bulb in all caſes waſtes away and conſumes, and it is the young bulbs produced from the fibres of the plant that ſpring out from it which are ever fit for uſe. I dug up ſeveral of theſe ſeedling potatoes in the ſecond and third years of their growth, in the months of June and July, and could obſerve no difference between the progreſs theſe plants had then made, and their whole œconomy, and others of the ſame kind planted in the uſual way. The bulb planted as ſeed waſted away nearly at the ſame period in both

caſes,

cafes, and the fibres from which the feeds originate began in both to appear about the fame time, and they feemed to be in every refpect alike.

The earlinefs of potatoes, *cæteris paribus*, feems to depend upon the nature of the kind planted, more than any other circumftance. Two kinds of potatoes planted at the fame time, and upon the fame foil, fhall differ fo much from one another in this refpect, that the bulbs of one kind will be fully formed and of confiderable magnitude, before the umbilical fibres of the other have begun to fpring forth, exactly analagous to what happens with regard to early and late kinds of peafe. It is poffible, however, that in potatoes of *the fame kind* thefe fruit-bearing fibres may fpring forth fomewhat earlier in very vigorous plants than in fuch as are more weakly; but I am ignorant if ever this fact has been hitherto afcertained. It is well known that rather the reverfe of this happens with peafe and beans, as the moft weakly plants (to a certain degree) of thefe claffes of vegetables come earlier than fuch as are extremely luxuriant.

[In the year 1779, I obtained from London a potatoe under the name of the early Henley potatoe; the bulb is a dirty white; form very round, not deeply indented at the eyes; fkin fmooth and fine,

fine, but not filky. The ftem low and dwarfifh, of a pale green colour; carries fcarce any bloffom, but the few flowers that appeared were of a pinkifh white, umbilical veffels very fhort, confiftence tending rather more to the vifcous than farinaceous; tafte fomewhat, though very flightly, fweetifh; the bulbs never of a large fize, feldom exceeding that of a large hen's egg.

By fome trials I made, which, not having been done with the accuracy I think neceffary, I do not fpecify particularly, it appeared that if this kind had been taken up at the beginning of Auguft, and at that time compared with the white kidney fort employed in my former experiments, the Henley fort would have afforded a crop *more than double* of what the other fort would have afforded. But had the crop been allowed to remain in the ground till the middle of October, the kidney fort would have produced more than twice as much as the Henley kind. My experiment was fufficiently accurate to allow me to reft fatisfied with this refult.

Hence it appears, that it would be equally bad œconomy in one who intended to lift his potatoes in the month of Auguft, to plant the *kidney* fort, if he could get the Henley, as it would be for him who intended to let them remain in the ground till October,

October, to plant the *Henley*, if he could get the *kidney* potatoe; for in both cafes he would only reap half the crop he might have done, had he made a judicious choice of feeds. The bulbs too of the Henley potatoes, at that early feafon, are much better to the tafte than thofe of the other kind, although the cafe is afterwards in fome meafure reverfed.

[This is one proof of the very great benefits that would accrue to agriculture, by an accurate experimental attention to circumftances.]

§. IV.

It has been alledged that potatoes, which have been long propagated by means of *bulbs*, lofe in time their generative quality, fo as to become much lefs prolific than at firft; and it is afferted that thofe bulbs which have been lately obtained from feeds are much more prolific, and confequently much more profitable for being employed as plants than others: but this opinion likewife I am afraid has been adopted without fufficient examination. I attended particularly to this circumftance in my own experiment, and could not obferve the fmalleft indication of fuperior prolificacy, in thofe raifed from feeds, but rather the reverfe.

That potatoes do *not* degenerate in point of prolificacy, in consequence of being long propagated in the usual way, seems to be confirmed by the general experience of all Europe. It is now about a hundred years since the potatoe was pretty generally cultivated in Ireland, and it has been very universally cultivated in Britain for fifty years past, and all that have been reared in it since their first introduction two hundred years ago, a *very few* of late only excepted, have been propagated from bulbs only; so that if they had declined in point of prolificacy, the degeneracy should in this time have been very apparent. Nothing of that kind however was ever remarked, nor any insinuation of that sort thrown out, till the discovery of rearing potatoes from seed was made, when it was for the first time heard of. There are many persons now living who have been in the constant practice of rearing potatoes for thirty or forty years; and notwithstanding the general tendency that mankind have to dispraise the present, when compared with past times, yet none of them have given the smallest hint of degeneracy in this respect. And I am persuaded, when it comes to be fully enquired into, it will be found that this is merely a groundless notion, that has originated from the partial fondness of those who first propagated this plant from seed, in favour of their new discovery.

PAPER

PAPER SIXTH.

THE DOCTRINE OF SEMINAL VARIETIES CONSIDERED.

§. I.

IT has alſo been ſaid, that by raiſing potatoes from ſeeds many new and valuable kinds may be obtained. An opinion of obtaining new varieties of plants by propagating them from ſeeds, ſo univerſally prevails among naturaliſts, and it had been ſo poſitively aſſerted as to potatoes, that I entertained no doubt about this matter, and waited with ſome degree of impatience till the time of taking them up arrived, to ſee what new varieties I ſhould thus obtain; but to my great ſurpriſe, and no ſmall diſappointment, I found no *new* kinds among my ſeedlings. There were indeed among them three or four varieties, but not one of them different from ſome of the kinds I had before; and as the ſeeds were picked up at random from a field in which all the different kinds had been intermixed promiſcuouſly, I think there is great reaſon to ſuppoſe that ſome of the ſeeds of the different kinds had been promiſcuouſly gathered, to which I attributed the varieties among my ſeedlings. This I mention however only as a ground for ſcepticiſm on this head, and not as a proof. As I did not at the time doubt of the fact, I took no precautions to have it ascertained. But this I shall endeavour to do on some future occasion.

My disappointment in this instance, however, made me examine with a greater degree of attention than I had ever before done, the whole popular doctrine of what is called *seminal varieties* among plants; when I found from other experiments I had formerly made, and observations I could recollect, that there is great reason to suspect that the received doctrine on that head is only a popular error. The following facts seem to confirm this opinion.

The phrase *seminal variety* has been adopted by botanical gardeners, and philosophical botanists, to denote certain smaller variations that are observable among plants of the same kind, than they have been able to find marks for discriminating. Plants by them are arranged into classes, which are further divided into *genera*, and these again into *species*. *Seminal varieties* denote different kinds that are all reducible to the same species, and they have obtained their name because it was supposed that they differed from each other only in some small peculiarities that were accidentally obtained from seeds, and that of course plants raised from seeds were apt perpetually to afford new accidental varieties of this

sort.

ſort. Being thus ſuppoſed to be perpetually liable to new changes, theſe ſeminal varieties have been totally excluded from every botanical ſyſtem of claſſification. As it is ſuppoſed that all the different kinds of potatoes uſually cultivated in Europe belong to *one* ſpecies, and that the different kinds are only accidental varieties that have been caſually obtained from ſeeds, it was very natural to ſuppoſe that new varieties of the ſame kind would ariſe from ſeeds whenever they ſhould be ſown.

According to the ſame ſyſtem, all the different kinds of cabbages, of turnips, of garden peaſe and beans, &c. are only *ſeminal varieties*, which, having been produced by accident, may be in like manner again varied by accident; yet that this is not the caſe, ſeems to be proved beyond diſpute by experience; for every gardener knows, that if he be at due pains in ſaving the ſeeds of his cabbages, &c. the ſame kind may be propagated for any length of time without the ſmalleſt variation: experience even goes farther, as it proves that theſe varieties may not only be kept diſtinct as long as you pleaſe, but that they may alſo, *in ſome caſes*, be mixed and adulterated almoſt at pleaſure; and that even theſe adulterated varieties can be made to continue of the ſame kind without any variation, as long as you ſhall

chooſe

choose to cultivate them, by a due degree of attention and care. I shall beg leave here to mention a few common and well-known facts, in confirmation of these remarks.

There are two kinds of cabbages very obviously distinguishable from each other, the *red* and the *white*. It is well known that if either of these kinds be cultivated in a district where none of the other is raised, no plants but of that kind will ever be obtained from the seeds there produced. No person who inhabits a part of the country where red cabbages are never brought to seed, ever saw a red cabbage plant raised from seeds of his own saving, nor the reverse. But it is also a fact equally well known, that if both white and red cabbages are reared in the same district, it is impossible, without very great precautions indeed, to keep the two kinds distinct, if an attempt be made to rear them from seeds of their own saving. The plants raised from seeds of the white cabbages become in this case, if nearly an equal proportion of each be saved, tinged with red veins, and those of the red become in some degree white, so that nothing but a mongrel breed, neither true red nor true white, can be obtained. If the proportion of one kind far exceed the proportion of the other, the kind of which there are feweſt ſoon becomes ſo much adulterated as to be ſcarcely in any reſpect different from the other; and thoſe who wiſh to have the leſſer quantity of a true kind are under the neceſſity of obtaining ſeeds from another diſtrict, where they are chiefly cultivated. In this manner thoſe of the ſouth of Scotland muſt obtain red cabbage ſeeds from Aberdeenſhire, and thoſe of Aberdeen muſt import their white cabbage ſeeds from elſewhere, if they expect to have them of a true kind. Phenomena exactly ſimilar to theſe occur in cultivating the red and white beet, the ſeeds of which always produce a mixed kind, unleſs they have been ſaved with great care.

A diſciple of Linnæus will find himſelf at no loſs to account for theſe phenomena, by drawing a parallel between the forementioned caſes, and the changes that are produced among the animal creation by an intermixture of different breeds of the ſame kind, which invariably produce a mongrel breed, participating of the qualities of both the parent ſtocks. I am fully ſenſible, however, of the danger of being miſled by ſuch general analogies in caſes of this nature, and ſhould not have relied upon that ſpecies of reaſoning, had I not been able to produce at leaſt *one* deciſive experiment on this ſubject: though I ſhall have occaſion to ſhew that

the

the rule is by no means ſo general as ſome, who rely on theory only, may be diſpoſed to believe.

Experiment Twelfth.

Among all the varieties of the turnip tribe, the yellow is the moſt remarkable, becauſe its colour is not confined to that part of its ſkin which appears above ground, as in moſt other kinds, but affects not only the whole of the ſkin, but the fleſh alſo. It is by much the ſweeteſt and firmeſt of all the turnip tribe; and inſtead of being injured by the winter's froſt, it is in ſweetneſs of taſte, and tenderneſs of conſiſtence, improved by it. On account of theſe qualities, it is highly valued for the table wherever it is known; but as it never attains to ſuch a ſize as the large green-topt field turnip, and is of a conſiſtence rather too firm for cattle whoſe teeth are tender, it occurred to me that if a mongrel breed could be obtained between theſe two kinds, it would anſwer extremely well for feeding cattle; and as the experiment could eaſily be made without trouble or expence, I reſolved to try if ſuch a kind could thus be obtained. With this view, adopting the principles of the ſexual ſyſtem of Linnæus, many years ago I planted ſome yellow turnips of a true kind for ſeed, and cloſe beſide them on both ſides I planted ſome green-topt turnips. In that ſituation they were allowed to flower and to perfect their ſeeds; and as care had been taken to prevent their flower-ſtems from intermixing with each other, the green-topt turnips were all taken away, and the ſeeds of the yellow kind were beat out by themſelves. Theſe ſeeds were ſown next ſeaſon, and produced a crop of yellow turnips tinged with a greeniſh caſt above ground, the fleſh of which was neither ſo deep in the colour, nor ſo firm of conſiſtence, as the genuine yellow kind, and the ſize conſiderably larger. It was in every reſpect a *mongrel* breed, which produced its own kind without variation for ten or twelve years, that I continued to cultivate it and preserve it from any intermixture with other sorts. If it were not a very easy matter for any one who can have access to green and yellow turnips to repeat this experiment for his own satisfaction, I should have been somewhat shy to mention the fact; but as the experiment is so simple and easily made, I presume any one who doubts it will take the trouble to satisfy himself experimentally about it. The same may be done with white and red cabbages or beets, where these more readily occur.

This experiment, at the same time that it seems in this case to confirm Linnæus's sexual system of the procreation of plants, in a still stronger degree seems to strike at the root of the doctrine of *seminal varieties*, in the sense it has been usually understood to bear, while it pretty clearly accounts for the origin of that doctrine. Men have observed, that in certain cases new varieties of plants, which had not been observed before, have been obtained from seeds. This might naturally happen where several varieties of the same kind of plants were allowed to flower, and to ripen their seeds together. The mongrel breed, especially among the gaudy tribe of flowers, sometimes might possess beauties that did not belong to either of the parent kinds, which would not fail to make them be particularly taken notice of. In this way, before any idea was entertained of any kind of sexual system in plants, this cause of the obvious change produced could not be adverted to, and it would be accounted an accidental anomalous seminal variety: and having observed this fact in several cases, it might be accounted a general law of nature, no pains having been taken to overturn it by opposing facts equally obvious at least. For the experience of the kitchen gar-

dener,

dener, who propagates without any variation of kinds, for centuries together, the different varieties of beans and peaſe ought certainly, if adverted to, to have excited a doubt at leaſt of the univerſality of the rule.

In the ſame manner that I obtained, as above, a new variety of turnips, it might poſſibly happen that ſome new varieties of potatoes might in ſome caſes be obtained from ſeeds. For if the varieties of this claſs of plants naturally admit of mixture with each other, (which however I doubt) and if ſeveral kinds of potatoes ſhould happen to grow in a field mixed together, two or more of theſe ſorts blended together might produce a mongrel breed, participating of the qualities of the parent ſtocks; but as to new varieties, obtained from ſeeds gathered from *ſolitary* plants, I have met with no fact that gives reaſon to expect them.

The only caſe that occurs to me juſt now as ſeeming to confirm the doctrine of ſeminal varieties, is that of fruit tree, which I only mention that it may not ſeem to be overlooked. But I have met with no facts that tend to prove that the generally received notion on this head is *not* hypothetical, and have found ſome that ſeem to prove that it *is*. It is in general very confidently aſſerted, and as generally believed becauſe of that aſſertion, that the ſeeds of a grafted apple do not produce trees bearing fruit of the ſame kind with themſelves. This is ſo generally believed in Europe, that I have never heard of an experiment being made with a view to prove it: but that very good apples are produced from ſeeds without grafting, is certain by the practice in America, where that operation is entirely laid aſide, and where the ſame kinds of apples are frequently found on different trees in the ſame way as among the grafted trees in Europe; though doubtleſs, where the ſeeds are allowed to ſow themſelves at random, there muſt be much lefs certainty, than where the practice of ingrafting prevails. Yet even here, on this very fubject of apple-trees, we meet with one fact that ftrongly oppofes the doctrine in queftion. There are two kinds of apples that have been ufually employed as ftocks on which others have been grafted. One of thefe is called *free* ftocks, and the other *crab* ftocks properly fo named. The plants of thefe two kinds are eafily diftinguifhable from each other by an experienced eye, and are well known by every nurfery-man. The curiofity is, that in fpite of the doctrine of feminal varieties, fo firmly believed by every gardener, the feeds of them may be bought as *diftinct kinds* in every feedfman's fhop, *and always produce plants according to their kind*, wherever the feedfman has been honeft. I need make no remarks on this fact.

Were I not afraid of tiring the reader, I could give many other inftances of plants, which contain many varieties that rife not above the clafs of *feminals*, even among trees and fhrubs, which *invariably* propagate by feeds their own kind; but fhall at prefent confine myfelf to the wild or dog rofe, many kinds of which produce a fruit called hips, which in the autumn are very beautiful. I have frequently gathered the feeds of particular kinds of thefe on account of their beauty, and fowed them; nor did I ever in any inftance know them to differ in any refpect from the parent plant.

§. II.

From what has been already faid on this fubject, the reader will clearly perceive, that although fome claffes of plants are certainly affected by being allowed to fructify by the fide of particular varieties of their own clafs; yet that there are others whofe

varieties

varieties are ſo diſtinct, that they ſeem to refuſe all ſort of intermixture with each other in any circumſtances whatever, and continue invariably to propagate their own kind by ſeed without any ſort of change whatever. This is obviouſly the caſe in all the varieties that are yet known of the pea tribe. Not only do the *grey*, the *white*, and the *green*, continue to produce their like without any variation, although they ſhould be reared together in the moſt promiſcuous manner, but even the leſſer varieties of each of theſe ſorts keep all their diſtinguiſhing properties without the ſmalleſt appearance of adulteration. A ſingle charlton pea that grew among a whole field of marrow-fats, if carefully ſaved, would produce next year a charlton pea of as true a kind as it would have done had it grown perfectly detached from all others; and the ſame thing is obſervable with regard to all the other kinds of garden peaſe. In like manner ſweet-ſcented peaſe, which conſiſt of four ſorts, the white, the purple, the painted lady, and the ſcarlet, though ſown promiſcuouſly, continue each to produce its own kind without any variation. All the kinds of beans poſſeſs the ſame property; as do alſo lupines, kidney-beans, and many other claſſes of plants. Hence it is obvious, that conſidered in this point of view, vegetables may b divided into at leaſt *two* general claſſes.

One

One claſs containing all ſuch plants as admit of a promiſcuous procreation, if the phraſe may be admitted, ſo as thus to produce a mongrel breed, as is particularly obvious in all the tribe of *Braſſica*. The other claſs, containing ſuch plants as do not admit of intermixture among the different varieties by procreation, each of which continually retains its diſtinguiſhing peculiarities, though reared in the neighbourhood of other varieties of the ſame claſs. This is particularly the caſe with all the pea tribe, and perhaps with all the papileonaceous claſs; though I dare not venture to ſay, that this rule would prove general. As this diſtinction has not, that I know of, hitherto been taken notice of, naturaliſts have not thought of aſſigning to each order of plants the claſs they ought to be placed under in that reſpect. Whether the *ſolanum* belongs to the one or the other, I cannot tell, and therefore cannot ſay whether any new varieties of potatoes may be expected from potatoes or not.

Conſidered in this point of view, there are probably other claſſes of plants that might be eſtabliſhed, with regard to which profeſſed floriſts may perhaps know ſomething. As I have no pretenſions to that name, I mention it here only to bring it under the view of thoſe who have opportunities

of

of inveſtigating ſuch queſtions. Some plants, when firſt raiſed from ſeeds, are ſaid to produce flowers of one colour only, which afterwards *break* as it is ſtiled, and become ſtriped, as the tulip. Others, though originally variegated, are ſaid in time to become plain, and afterwards retain that plainneſs, as the carnation. Whether theſe things are actually ſo, I do not take upon me to ſay, becauſe I know many things of this nature are vaguely and confidently aſſerted. But if they do exiſt, it might be of uſe, for ſome purpoſes, to aſcertain the plants that are reducible to the one or the other of theſe claſſes.

Upon the whole, with reſpect to the rearing of potatoes from ſeeds, I think we may ſafely conclude, that if this practice will ever be productive of any benefits to ſociety, theſe advantages have not yet been diſcovered.

PAPER

PAPER SEVENTH.

OF THE DISEASE CALLED THE CURL IN POTATOES, AND SOME OTHER PECULIARITIES OBSERVABLE WITH REGARD TO THIS PLANT.

§. I.

POTATOES are subjected to one particular disease, the *curl*, which it would be improper for me altogether to omit; and though I can say little *positive* as to the cause of this disorder, a good deal may be said on the *negative* side of the question; which, as it may possibly lead to future researches, shall here be added.

The only thing that seems to be positively certain with regard to this disorder is, that it was scarcely, if at all, known till very lately; and in particular that it was not known in the northern parts of this island till a very few years ago, (I myself had never seen it when the former parts of this treatise were written) when, there is great reason to believe, it was introduced by means of seed potatoes imported from the South country: and it is at this moment much less frequent in the Northern and remote parts, than in the Southern and more commercial districts of this island.

From

From this ſingle fact, ſeveral corollaries are deducible, which entirely overturn ſome theories that have been formed to account for this diſeaſe. It has been ſuppoſed, that nature, ſated as it were by having long produced this plant in a climate that was not deemed congenial to it, had become ſo far exhauſted, as to occaſion this diſeaſe. But if ſo, the more Northern parts of this iſland, where the climate is the moſt unfavourable, ſhould have been ſooneſt affected by it. It has been further ſuppoſed, that potatoes that are froſt-bitten, (the bulbs are here meant) before they are properly houſed, occaſion this diſeaſe in the plants produced from them; but the curl is leaſt known in thoſe diſtricts where the potatoes are moſt liable to this accident. It has been ſaid, that potatoes which are taken up before they attained perfect maturity, infallibly became affected with this diſeaſe: but in ſome cold moſſy ſoils, and expoſed ſituations, where the potatoes muſt often be taken up while they are yet in a ſtate of vigorous vegetation, this diſeaſe is ſcarcely known. It will not be imagined I mean to inſinuate that ſuch ſituations would preſerve from the diſeaſe, when once introduced in other places around; for that, without doubt, is not the caſe.

It has been further alledged, that ſuch plants as have been lately obtained from ſeeds, properly ſo

called,

called, are entirely free from the *curl.* But I have a very ſtriking proof before my eyes at this moment, that this is not the caſe. A large field, which was planted with potatoes the third year from the ſeed, has more than one half the plants curled; while another field near it, raiſed from potatoes that never were, that we know of, produced from ſeeds in this country, has ſcarcely one curled plant in the whole.

It is by ſome alledged, that the ſoil or ſeaſon occaſion the diſeaſe: but that this is not the caſe, is alſo plain from the ſtate of a field now in my neighbourhood. Several ridges in it, that were planted with potatoes obtained from one particular perſon, are very much curled, while the remainder of the field, which was planted with potatoes of the ſame ſort, obtained from a different perſon, is not at all infeſted with that diſeaſe. In this caſe the ſoil and climate were the ſame, (for the ridges were intermixed with the others) but the crop extremely different.

In the laſt example it is plain, that the diſeaſe depends entirely on the nature of the ſeed; and it ſeems to me highly probable that the curl in potatoes, like ſome hereditary diſeaſes among animals, if once introduced, vitiates the prolific ſtamina, ſo as to be perpetuated as long as the infected breed

continues

continues to produce others. But this is hazarded merely as a conjecture. Are there any facts sufficiently authenticated, which prove that a plant once known to be infected with this disease, invariably produces diseased plants? Or is it for certain known, that in any case a curled potatoe has been produced from a plant that was perfectly sound, and not in the smallest degree affected with that disorder? Clear proofs by experiments made with the utmost caution are here wanted, and not the result of random observation.

In considering these questions, and comparing them with phenomena already known, it seems difficult to decide which way the probability lies. On the one hand, seeing this disease is, or at least most certainly *was*, entirely unknown in many large districts where potatoes have been long cultivated, it would seem to favour the opinion that it only originated from infected seed:—on the other hand, it is asserted, as a well-known fact, that fields which have been planted with seed that was in the former year very little if at all affected, have been known to produce plants the succeeding year, almost entirely of the curled sort. Should this fact be proved, it would seem to favour the opposite hypothesis; but even here, we shall soon see reason to judge cautiously.

Infectious diseases in the animal creation may be communicated in two ways. One class of disorders can be communicated only by descent from parent to child, and can never be communicated by juxta position: another class of diseases can be communicated merely by juxta position, and not at all by descent of blood. Does such a distinction as this prevail among vegetables? Has any experiment been recorded, which proves that any particular disease among plants can be conveyed by juxta position only? Is not the smut in wheat of this class? May it not extend to others? May not a few infected potatoes in this way, if mixed in a large heap, like a subtile leavening principle, infect the whole? Experiments are here much wanted.

But I am far from alledging, that, though it were an established fact that potatoes had been known to be much more infected with the curl, than those from whence the seed was obtained had been in the former year, it would either prove that the disease might originate from other causes than contaminated seed, or make it certain that besides direct descent, the disorder could be communicated merely by juxta position. For though it should so happen, that the disorder could be communicated only by direct descent, the fact above-mentioned might poſſibly happen in this way. It is well known that potatoes affected with the *curl*, ſeldom produce bulbs of a ſize nearly equal to thoſe of the ſame kind that are ſound. This being the caſe, ſhould one man, from among the heap of thoſe potatoes which were only in a ſmall proportion affected with the curl, ſelect only the ſmall potatoes for *ſeed*, and otherwiſe diſpoſe of the large ones, it is obvious he muſt thus preſerve almoſt the whole of the curled ſort for ſeed, and very few of the ſound; and the potatoes raiſed from this ſeed muſt of courſe be much more infected with the curl, than the parent potatoes were, from which the ſeeds were obtained.

On the other hand, ſhould another man pick out from the ſame parent ſtock only the very largeſt bulbs he could find, he would probably have ſcarcely *one* curled plant in the whole field. Thus might the ſeed from the ſame heap prove to be, in one field entirely free from the infection, and in the other altogether diſeaſed. Men are in general ſo careleſs in obſerving facts of the nature here alluded to, that we frequently meet with inexplicable phenomena like what we have here ſuppoſed. But till theſe particulars be fully inveſtigated, it would ſeem to be, from many conſiderations, the ſafeſt plan to ſelect only the *largeſt* bulbs for plants: for as there is reaſon to think the diſeaſe proceeds in all caſes from the ſeeds planted, and as the infection muſt certainly be leſs virulent in the plants that have been leaſt previouſly infected with it than others, and as no large potatoes are produced by plants which are ſtrongly affected by this diſeaſe, theſe ought ſurely to be choſen for plants in preference to others. This I know, that I never yet have obſerved a curled potatoe among thoſe I have obtained from *large* potatoes planted whole.

Should it be found that this diſorder, like the ſmall-pox or meaſles among mankind, can be communicated by juxta poſition only; in that caſe it would be prudent to ſelect the large bulbs at the time of taking them out of the ground, putting them all in one heap then, without ever letting them touch the ſmall, and thus keep them entirely ſeparate. By theſe cautions, accompanied with roguing the potatoes as ſoon as they are fairly above ground, (that is, pulling out all thoſe that ſhewed the ſmalleſt ſymptom of this diſeaſe) it ſeems to me probable it might be in a great meaſure, if not entirely, avoided.

Some think the diſeaſe is produced by allowing the potatoes to be a little touched by froſt during the winter. Certain it is, that a potatoe never can

be

be in any respect benefited by frost, and therefore cannot be too carefully guarded against it; but from the facts already stated, it does not seem to me at all probable that ever the disease originates from that cause.]

Besides the above, there are many other particulars relative to the culture of this useful plant, that require to be elucidated, some of which are as under.

It is to be observed, that although the same kind of potatoe be planted in two different fields, the produce of the one often proves to be of a much more meally kind of potatoe than that of the other; and it has been in like manner remarked, that the potatoes of one year's growth are in general more watery, or the reverse, than those of another season. The causes of these peculiarities do not seem to be as yet fully known, though, as is usual, many things are vaguely asserted on this head, with a sufficient degree of confidence.

It is generally believed that a dry soil, or a dry season, necessarily produce the driest potatoes, and indeed it is so natural to expect, *a priori*, that this should be the case, that it is not surprising if men should not be difficult about admitting the fact. I find good reason, however, to suspect that these opinions are not altogether well-founded. The year 1775 was the drieſt and warmeſt ſeaſon that has been known in Scotland within the memory of man, yet the potatoes of that year's crop were watery almoſt to a proverb: on the other hand the potatoes of crop 1777, although it was a remakably rainy ſeaſon, were as dry and meally at leaſt as is common, and much more ſo than in the year 1775. It deſerves alſo to be remarked, that the crop of 1775 was almoſt double in quantity to that of 1777. Hence a dry ſeaſon would ſeem to augment the produce, though it does not for certain in all caſes improve the quality of this crop.

The year 1774 was one of the coldeſt and moſt rainy, that has perhaps been known in Scotland. *Quere*—Could that have any effect on the produce of the enſuing ſeaſon? If it had, the potatoes of crop 1778 ſhould be more watery than uſual; for the year 1777 was almoſt as cold and rainy as 1774. Obſerve if this ſhall happen.

[N. B. The potatoes of 1778 were not more watery than uſual, therefore this conjecture does not ſeem to be well founded. The year 1782 was the coldeſt and wetteſt ſeaſon ever known by any man alive: but the potatoes were almoſt entirely deſtroyed by froſt in Aberdeenſhire, before they were taken

taken up; and my attention was so much engaged at that time with more interesting concerns, that the present subject of investigation never occurred, so that I made no remarks on that crop.]

If a dry season does not necessarily insure meally potatoes, so neither does a dry soil always and necessarily produce that effect. I have frequently seen the same kind of potatoes, and of the same year's produce, reared in two different places; the one of them in a soil which was naturally inclined to dampness, which were much freer and more meally than the others which were reared on a drier and sharper soil. This I have so often remarked, that I cannot be mistaken as to the fact. I have, no doubt, as often observed that the case has been reversed. I believe it will be also found to be a fact, that potatoes which are raised in those districts of the country, where the soil is of a hot sandy nature, are *usually* more free and tender than those which are reared in countries where the soil in general is cold and damp. Such seemingly contradictory phenomena as those abovementioned are not uncommon in agriculture, and often give rise to false opinions, which are followed by erroneous practice. In order to discover the cause of these seeming contradictions, conjectures may be freely hazarded, if they are delivered only as *conjectures*, not to influence our reaſoning, but to direct the attention towards proper objects of enquiry and experiment. It is not even neceſſary that theſe conjectures ſhould be founded on any facts already known; it is enough if they point towards probabilities, that may be confirmed or refuted by future obſervations. They ought not even to be conſidered by the perſon who hazards them as probabilities, which it imports his character to ſupport, becauſe this would warp his judgment, and pervert his reaſoning; but as mere random gropings in the dark, which, if they do not clearly diſcover what is the direct road to knowledge, will at leaſt in ſome caſes point out what is *not* the track to be purſued, and will at any rate ſooner diſcover it, than if we ſtood ſtill without exertions or obſervations of any ſort.

With theſe views I would hazard the following query;—Is the waterineſs or dryneſs of a crop of potatoes in any ſort affected by the degree of ripeneſs that the plants employed for ſeed may have attained in *the preceding ſeaſon?* That the maturity they have attained in *the ſeaſon that the potatoes are uſed*, does affect the quality of the potatoes, I conceive to be highly probable; and therefore potatoes, which, on account of the richneſs or other peculiarity of ſoil, continue in a ſtate of vegetation highly luxuriant, till they are nipped by froſt or checked

in

in their growth by other inclemencies of ſeaſon, have much leſs chance of being dry and meally, than others of the ſame ſort, which have attained their full growth before the coldneſs or inclemency of the weather checked them. The preſent queſtion, therefore, does not relate to this, but to the effects that ſuch unripe plants have upon thoſe produced from them next ſeaſon. If, upon examination, it ſhould be found that the due maturation of the plants employed as ſeed had any effect upon the quality of the future crop, it might help us to account for ſome of the foregoing phenomena; becauſe, in a country of various ſoils, it might accidentally happen, that the crop raiſed on a dampiſh ſoil was produced from ſeeds that had grown on a dry warm ſoil the preceding year, and had been ſufficiently ripened, or the reverſe; but in large diſtricts, where the ſoil is in general pretty much of one quality, either warm and dry, or cold and wet, the kind of interchange of ſeed here alluded to could not ſo readily take place.

But even if it ſhould be found that the maturity of the ſeeds affected the quality of the potatoes, it would not follow invariably that the ſeeds produced on early dry ſoils would be better than thoſe from later ſoils, becauſe it might ſome times happen, from local position, and other accidental circumstances, that the growth of the potatoes in the dry early soil might be checked by frosts many weeks before those on the other soil were affected; in consequence of which, the plants in the cold soil might attain to more perfect maturity, than those on the drier one. I mention this peculiarity, merely to shew how cautious the farmer ought to be in adopting general conclusions, without carefully attending to all the collateral circumstances that may affect his experiment. I shall only farther add on this head, that I had occasion to know well a dry warm spot of ground, on which the stems of the potatoes of crop 1776 were frost-bitten, at least *six weeks* before those on another spot at some miles distance from it, where the soil was naturally more cold and damp, were in the smallest degree affected by it. It likewise so happened, that the potatoes raised on the first-mentioned spot in the year 1777, (their own frost-bitten* seed was employed) had such a peculiar acrid and bitterish taste, as to be hardly at all eatable; while those in the colder place of that crop had nothing of that unusual taste. Whether this diversity was occasioned by the circumstance here

* Observe, the term *frost-bitten* is here applied to the stems only, and not to the bulbs. The stems were so much hurt by the frost as to turn black and decay, but the bulbs were taken up before the frost had been sufficiently intense to hurt them.

alluded to, I do not take it on me to ſay. In matters of ſuch nice diſquiſition as the preſent, many facts obtained by very accurate obſervation are neceſſary, before any concluſion can be relied on.

The following accidental experiment, relating to the ſubject here in agitation, deſerves to have a place:—

Experiment Thirteenth.

In the year 1776, I planted with potatoes a ſmall plot in my garden; it accidently happened that the one half of it had been in cabbages the year before, and the other half in graſs, which, for the ſake of an experiment, had lain in that ſtate for three years. The ſoil was in every other reſpect the ſame. The whole was dug over in the month of April, ſome looſe mould having been ſhovelled up on the top of that part which had been in graſs, merely to cover the graſſy part of the ſod. It was all planted with the ſame kind of potatoe on the ſame day, and managed in every reſpect alike. None of it got any dung. The crop was in both places very good, and nearly equal in quantity; but it was remarked, as a ſingular peculiarity, that the potatoes which grew upon the part that had been in graſs were remarkably meally, whereas thoſe that grew upon the other diviſion were of a very ſoft and watery kind. The difference between them was ſo perceptible, that no perſon in the family but could have told at once if the ſervant by miſtake at any time brought the one kind inſtead of the other.

In this caſe, it is obvious, that the difference in quality was produced entirely by ſome peculiarity in the ſoil, and could

could neither be occaſioned by any defect in the ſeed, nor peculiarity of weather; and on this occaſion I imagined I had diſcovered a circumſtance that had hitherto baffled all my reſearches: for I thought it next to certain that the ſuperior mealineſs of the one part of the potatoes in this experiment, was occaſioned by the ground on which they were planted having been broke up directly from graſs; and although I could aſſign no probable reaſon why this ſhould be ſo, yet as no other difference between them was obſervable, I reſolved to repeat the experiment, to ſee if the ſame phenomena regularly occurred. This produced the following trial:

Experiment Fourteenth.

In the year 1777, I made choice of another patch of ground, one half of which had been in culture many years, and the other half was in graſs three years old. Both of theſe were dug over in the month of April, exactly in the ſame manner as in the foregoing experiment, and were planted as before, with one kind of potatoes on the ſame day. In every reſpect theſe were treated, as nearly as poſſible, in the ſame way with thoſe in the laſt experiment. But when they were taken up at the proper ſeaſon, to my great mortification, I found that no ſenſible difference could be obſerved in the friability of the potatoes obtained from the one or the other diviſion. The reader will alſo pleaſe to recollect, that experiments ſecond and third were made upon ground in like manner newly broke up from graſs; but neither were the potatoes that were produced upon it, although it was a dry, ſharp, thin ſoil, not at all remarkable for their dryneſs or mealineſs: they were even much inferior in this reſpect to thoſe which were obtained from both the diviſions of the preſent experiment; although the ſoil was, in the laſt caſe, of a deeper and damper kind.

It

It is thus that knowledge frequently eludes the researches of the farmer, after he thinks he has with certainty attained it; but if he be diligent and unremitting in his pursuit, and never gives over, even when he seems to be thrown out in the chace, he will at length lay firm hold of this ever-changing Proteus, and force from him many important secrets exceedingly necessary to be known for the well-being of mankind.

CONCLUSION.

The reader cannot fail to have remarked, that the foregoing experiments and observations only tend to pave the way for an accurate set of experiments, to ascertain with some reasonable degree of precision, the soil, manures, and culture, that are best calculated to produce the largest and best crops of potatoes. Till the particulars above specified be fully ascertained, any attempt to prescribe the best and most advantageous mode of cultivating this valuable plant must be vain and nugatory, as perpetual contradictory facts would occur, which would involve the subject in the same doubt and obscurity as at present. Fully convinced of these things, my aim in this essay has been solely to elucidate some important previous questions. Little more indeed has been done, than to point at what is wanted for enabling us to go forward in a proper manner: and these

these imperfect hints are submitted to the public, in hopes of inducing others, who have better opportunities of making experiments than myself, to exert themselves in an effectual manner to ascertain those points that still are doubtful. I shall myself endeavour, as far as circumstances permit, still to go forward in this tract, and am not without hopes, that in time I may be enabled to speak with some degree of firmness, concerning the modes of culture that are well adapted to insure great and profitable crops of this most valuable plant. At present, I rather chuse to decline entering on that branch of the subject.

N. B. *As there is much diversity in weights and measures in different parts of this country, readers are often greatly embarrassed for want of knowing the exact amount of those that are mentioned in experimental essays. To avoid that inconveniency on the present occasion, the reader is desired to take notice, that, unless where it is specially mentioned to be otherwise, throughout the whole of the preceding essay, an* acre *means an exact statute English acre of 4840 square yards. A* pound *means an avoirdupoise pound of 16 ounces, and a bushel 56 of these pounds, or half a hundred weight. Every reader, by the help of this information, may easily bring any weights or measures mentioned to the same standard that is used in his neighbourhood.*

The greatest part of the essay was written in the year 1778; a few observations having been since added, which are distinguished from the original essay by being included within crotchets [thus.]

Article II.

An Essay on the most practicable method of fixing an equitable Commutation for Tithes in general throughout the Kingdom.

TO THE PRESIDENT, VICE-PRESIDENTS, AND MEMBERS OF THE BATH AGRICULTURE SOCIETY.

THE judicious proposal of the Bath Agriculture Society, for an "Essay on the most practicable method of fixing an equitable commutation for tithes in general throughout the kingdom," reflects great honour on the institution.

It is, I believe, universally acknowledged, that tithes are a great discouragement to agriculture. They are inconvenient and vexatious to the husbandman, and operate as an impolitic tax upon industry. The clergyman too frequently finds them troublesome and precarious; his expences in collecting are a considerable drawback from their value, and his just rights are with difficulty secured: he is too often obliged to submit to imposition, or be embroiled with his parishioners in disputes and litigations, no less irksome to his feelings than prejudicial to his interest, and tending to prevent those good effects which his precepts should produce.

The

The writer of this essay has frequently been consulted about the value of tithes, and that of the land out of which they issue; as also on the comparative value of one to the other. It is from observation, and reflection, grounded on experience, that these hints are submitted to the Society. Had the author sufficient leisure and abilities for entering at large into the present establishment for the maintenance of the clergy, and for stating how peculiarly hard it bears on the landholder, when compared with the merchant, the manufacturer, and the artisan; such a discussion, however worthy the attention of parliament, is not the object of this essay.

In the practicable method to be pointed out, it appears indispensable, that a fair, full, and permanent equivalent should arise out of the same property, and be defrayed by the same order of men, as pay tithes at the present day: the more any proposed scheme deviates from this principle, the less practicable will it be found.

Land for Tithes.

A commutation of tithes for land has many advocates, and some very able opponents. The Lord Bishop of Salisbury, in a late excellent charge to his clergy, has amongst other important matters shewn, with

with great ſtrength of argument, that ſuch a commutation is by no means eligible. The habits of life in which the clergy are educated, and the important office they fill, are ill-ſuited to the occupation of a farmer. The expence requiſite to ſtock a farm would, to many, be a ſerious objection. If we conſider the land ſo taken only as property to be let, the moſt deſirable circumſtance would be for it to lie compact, and as near the buildings as poſſible. In extenſive pariſhes, where there are numerous ſmall eſtates, this object is unattainable. The equivalent in land muſt in ſuch caſes lie in very ſmall parcels, exceedingly diſperſed, and be difficult to let to proper tenants at a fair value. A balance muſt be ſtruck upon each eſtate, and fences be raiſed at a a great expence. Such parcels as would be eligible for the rector to receive, the landholder cannot always give, without deranging the general œconomy of his farm. Even in thoſe pariſhes where the property may be given and received with conveniency, and let to one tenant, he is liable to misfortunes, and failures, which would render the clergyman's ſupport more precarious than on the preſent eſtabliſhment. The knowledge of ſoils and their uſes, requiſite for framing covenants for the preſervation and proper management of landed property, will frequently be wanting in the clergy. And the probability

bability that a fucceffor will find the land neglected or exhaufted, the fences deftroyed, and the buildings in ruins, will not be doubted by fuch landowners as have declined to renew with their lifehold tenants, and fuffered their property to fall into hand.

Whoever has taken an active part in carrying inclofure bills into execution, where the land is exonerated from tithes, muft have found it a nice and difficult tafk to afcertain a proper equivalent. When the proportion of land to be given is fixed by parliament, it is too often done without fufficient information refpecting the circumftances of the property. Hence may be affigned a principal caufe why the real merits of inclofure bills are frequently depreciated. The great diverfity of foils, their different degrees of fertility, various ufes and products, and the different expences of cultivation, all operate on this proportion; and what may be deemed equitable on one eftate, is frequently injurious to the rector, or to the landholder, on another.

A very able writer* ftrenuoufly oppofes a commutation of land for tithes, and that chiefly as being injurious to the landholder.

* See Obfervations, &c. refpecting Bills of Inclofure; and calculations fhewing the lofs fuftained wherever lands are given in lieu of tithes.——SANDFORD, Shrewfbury; and BEW, London.

"Let

"Let us ſuppoſe a farm of 150 acres, at 16s. per acre; rent 120l. The leaſt profit ſuch a farm ought to produce, in order to anſwer the various expences incident to it, is 300l. It ſhould be more; ſtate it however at that ſum, and the account will ſtand thus:

	£.
Rent, - - - - - - - -	120
Tithes - - - - - - - -	30
Remainder to anſwer every expence of pariſh rates, wages, houſekeeping, wear, tear, &c.	150
	300

"Here the value of the tithes is equal to a fourth of the rent; and I take this to be the leaſt proportion that the tithes and rent can bear to each other in any caſe whatſoever, except in rich grazing farms managed at a ſlight expence. Now that the rector may have a full equivalent of landed rent for theſe tithes, an allotment muſt be made worth 30l. a year tithe free. But the tenant pays 150l. per annum in rent and tithes, or 20 ſhillings per acre; conſequently the allotment muſt contain 30 acres, leaving the remaining 120 in the occupation of the tenant, who muſt continue to pay his uſual rent; for otherwiſe his landlord muſt ſuſtain a loſs. The account will then ſtand thus:

Tenant.	£.
Rent of 120 acres, - - -	120
Reſidue for general expences, - -	120
Produce, reckoning as before at the rate of 300l. upon 150 acres - -	240

Rector.	£.
Thirty acres let at 20 ſhillings per acre,	30

"Thus, by a commutation neither benefiting nor injuring landlord or rector, the tenant is reduced from 150 to 120 pounds, to ſupport nearly the ſame family, and defray within a trifle the ſame expences. This is an actual loſs to him of little leſs than 30l. per annum."

It muſt be acknowledged, that the above writer takes no notices of the expences of collecting tithes, and converting them into money. Servants' wages, horſes, carts, reparation of barns and other buildings, waſte, threſhing, marketing, &c. are conſiderable deductions. In ſtating a proportion in the rule of three, the young arithmetician is directed to reduce his firſt and third terms to the ſame denomination. It is equally reaſonable, if the expences of the cultivator are reckoned, that thoſe of the tithe-gatherer ſhould not be forgotten—that the letting value of land ſhould be compared with the letting value of tithes; not the rent of the one with the produce of the other.

Although this omiſſion does not invalidate the general principle; it enhances the comparative value of tithes to that of land. It is a very common omiſſion in calculations of this ſort, and may, in a work of merit, tend to miſlead; on which

account

account it is here taken notice of. The ſame author mathematically demonſtrates, that land cannot be given for tithes, in any caſe whatever, without injury to one or both of the parties concerned—and that the meaſure of that injury muſt be a ſum nearly equal to the profit which accrues to thoſe who afterwards occupy the tithe allotment. As this is a queſtion of conſiderable conſequence, it cannot be too nicely and impartially examined. Admitting the author's aſſertion to be *mathematically* true, it is alſo true that a given quantity of land, rented as part of a very large farm, is generally leſs valuable to its owner, and to the community, than when it forms part of a farm of more moderate ſize. If a commutation of land for tithes were only to take place on overgrown farms, neither the landlord nor the public would, perhaps, have any cauſe to object. But the conſequences of ſuch a commutation would be moſt ſeverely felt by a ſmall land-owner, who is obliged to keep a certain number of horſes to till his ground, whether he has a few acres more or leſs. The profits ariſing from his labour, even when in full employment, are barely ſufficient for a comfortable livelihood: take away a portion of his land, and, like the manufacturer who is obliged to ſtand ſtill for want of materials, he is in part deprived of the means of maintaining his family.

Through a refined taſte, or a miſtaken policy, the induſtrious occupiers of ſmall property, thoſe moſt uſeful ſubjects to the ſtate, have been already too much oppreſſed, and are in many places nearly extirpated. The good-natured reader will excuſe me, if I quit my ſubject for a moment to deplore the conſequences.

The Norman Conqueror, in the plenitude of his power, depopulated thirty villages for his pleaſure, which has left an indelible ſtain on his character. In the preſent enlightened age, the ſame ruinous policy is adopted, without remorſe or cenſure. If the homely habitation of induſtry grows cold and comfortleſs, avarice whiſpers that the expence of reparations may be ſaved, and the land be added to a neighbouring farm. If parks or pleaſure grounds are to be extended, whole villages are razed to the ground. Huſhed are the cheerful " ſounds of population," and the buſy footſtep is ſeen no more. The once comfortable, but now dejected inhabitants, are reduced to the hard neceſſity of earning a ſcanty morſel in the evening of life, by dint of labour beyond their declining ſtrength; and thus their grey hairs go down with ſorrow to the grave:—their beloved children, the comfort of their age, ill-brooking the idea of ſervitude, where they have ſeen better days, ſeek employment in the capital;—diſappoint-

ment

ment and penury enſue:—the mounds of virtue are now broken, and the ruddy bloom of health exchanged for diſeaſe and infamy:—our ſtreets become crouded with ruined innocence; and our priſons with wretched and deſperate malefactors!

Where are the benefits to compenſate for this maſs of evil? The property which maintained ten or twenty families in comfort, is now converted into a ſingle farm. When a tenant is wanted, there are but few competitors. If he fails, the leſs is a ſevere one. The landlord may conſole, and perhaps reimburſe himſelf by ſeizing on the farmer's ſtock; but his feelings are not to be envied; the reputation of his farm is thereby leſſened, and the difficulty of procuring a tenant increaſed. The man who can afford to ſtock ſuch a farm, can probably live on the intereſt of his money. He will not embark without a probability of large profit; and that without taking a laborious part. The labour and attention of ſervants and workmen are more expenſive and leſs effectual than that of a ſmall farmer, who eaſily ſuperintends his buſineſs, and, with the aſſiſtance of his children, tills his own ground. The graduated ſcale of property being broken, and no medium left between the overgrown farmer and neglected cottager, the ſinews of induſtry become languid. The poor man has nothing to look up to:—No motive for a laudable pride—no incentive to ſuperior induſtry. The pariſh poor are his aſſociates; and he obſerves that, when age or want overtakes them, the diſſolute and the worthleſs are indiſcriminately, and *equally*, reli eevdwith the worthy and the induſtrious.—He therefore literally takes no thought for the morrow.—The produce of his labour is ſpent without reſerve; and his wretched family entailed on the pariſh:—the land becomes loaded with enormous poor-rates, and its owner, after all his ſchemes of aggrandizement, wonders to find its value decreaſed. Such are the evils which prudence would have foreſeen; and a humane attention to the rights and intereſts of mankind ſhould have prevented. But to return.——On the moſt mature conſideration, I am fully convinced of the impracticability of fixing a *general* and *equitable* commutation of tithes for land throughout the kingdom.

COMPOSITIONS.

The fluctuating value of money, and the very ſmall proportion which moduſſes, or real compoſitions, made previous to the diſabling ſtatute 13 Elizabeth now bear to the value of tithes then compounded for, are convincing proofs that no pecuniary

cuniary payment can be fixed, without the greateſt probability of injuring poſterity. To remedy this evil, proviſion-rents have formerly been adopted. But a moment's conſideration will convince us that a commutation of tithes for proviſion-rents, or the produce of the land in a marketable ſtate, would be liable to much trouble and abuſe. It would not be eligible even for the farmer; nor can it be expected that the tithe-owner would ever conſent to it.

The worthy and learned Prelate before-mentioned, with equal judgment and philanthropy, recommends to his clergy to compound with their pariſhioners on *moderate* terms. Were this ſalutary advice univerſally adopted, it would be for the benefit both of the clergy and the laity. For it is well known that beſides the trouble of ſetting out tithes, and their numerous ill conſequences to the landholder and to ſociety, they are in collecting liable to waſte, injury, and additional expence: and that there is a very conſiderable loſs between the rector and the farmer, without benefit to the one or the other. But, alas! plain as this truth muſt appear to every man of experience, the imperfections of human nature are ſuch, that the parties concerned rarely agree on an equitable compoſition.

To prevent the effects of that partiality, which want of judgment or of candour too often occasions, the ingenious writer before quoted proposes, that the sum which each person shall pay in lieu of tithes should be fixed by two indifferent and skilful persons, with liberty to any of the parties to order a new valuation to be made once in every seven years: the expences to be equally borne by the rector and the parish.

The expence attending a measure of this sort, often reiterated, would be one considerable objection. Many improvements in husbandry are attended with heavy expences, and the return is frequently uncertain. A septennial reckoning with the tithe-owner may damp that spirit of industry, which an exoneration from tithes is meant to produce. A disagreeable anxiety would attend property often submitted to arbitration. And when we consider that men have been perverse enough to let their lands lie unsown, in order to deprive the rector of his tithe, we may take it for granted that there would not be wanting those, who, with an unworthy policy, would take every possible step to warp the judgments of the arbitrators, by depreciating the value of their tithes previous to such septennial, or any other regular valuation.

COMPOSITION TO VARY WITH THE VALUE OF MONEY.

The moſt unobjectionable commutation that occurs to me, is that of a money payment, chargeable on the occupiers of the land now titheable, but to vary with the value of money, in ſuch manner as for the ſame income to purchaſe the ſame quantity of the neceſſaries of life, in all times to come.

In order to accompliſh this end, it is propoſed, that a bill be brought into parliament, not to compel every pariſh to enter immediately on ſuch a meaſure, but to enable all parties, who are deſirous, to proceed on the buſineſs. A very ſudden and general change would neither be practicable, nor eligible. It muſt inevitably be a work of time; and ſhould be carried on rather from conviction than compulſion. The following hints may probably be of ſome uſe in framing the principal clauſes in a bill for that purpoſe.

MODE OF PROCEEDING.

That every thing may be tranſacted in as ſhort a manner, as openly, and at as little expence as poſſible, the Juſtices at their general Quarter-Seſſions of the peace, held for the ſame county, and at the neareſt diſtance from the pariſh where the tithes are propoſed to be commuted for, ſhould be enabled to authorife commiffioners to proceed on the bufinefs. Whoever has attended the paffing of private bills through parliament, muft have obferved, that although the allegations of a bill are proved before a Committee, with due care and folemnity, yet the real merits and moft material parts, are fometimes but imperfectly underftood, or attended to. At a general Quarter-Seffions, it is probable, that feveral of the magiftrates may be well acquainted with the merits and circumftances of the bufinefs, and the expence of attendance will not debar the parties from coming forward with the beft evidence that can be obtained.

Two months previous to an application to the Quarter-Seffions, a notice fhould be affixed on the principal door of the parifh church, for three Sundays during divine fervice, fetting forth that application will be made on the firft day of the next general Quarter Seffions of the peace, held at —— ——. And alfo an advertifement to the fame purport fhould be inferted in fome country news-paper, which circulates in that part of the county where the lands are fituate, in order that the juftices and non-refident parties may in proper time be apprifed of the bufinefs. That fuch notices have been given, fhould in the firft place be proved upon oath.

oath. The proportion of conſenting parties ſhould next be brought forward, and the reaſons given by thoſe who withhold their aſſent ſhould be particularly ſtated.

PROPORTION OF CONSENTS.

The parties intereſted in the tithes, or thoſe whoſe conſent ought to be obtained, are, the Biſhop of the dioceſe, the Patron and Incumbent—appropriators holding tithes, and impropriators, with their reſpective leſſees for long terms renewable or for lives. Whatever be the denomination of the parties intereſted, a general conſent ſhould, if poſſible, be obtained. But as this is not always to be expected, however meritorious the undertaking, it may, perhaps be thought that one diſſenting voice on the part of the tithe-owners ought not to negative the buſineſs. However, in all caſes where the patron is alſo incumbent, or where the tithes are a lay impropriator's freehold, or wherever the poſſeſſion and reverſion center in one and the ſame perſon, ſuch perſon's conſent ſhould be indiſpenſably neceſſary.

On the part of the land-owners, the conſent of the proprietors of three-fourths of the property in quantity or value ſhould be obtained, previous to ſuch application to the Quarter-Seſſions. But no

perſon

perſon who conſents or diſſents as dioceſan, patron, incumbent, or tithe-owner, ſhould have any vote as land-owner alſo. If any oppoſition be made to the meaſure, the parties oppoſing ſhould be heard by themſelves or counſel; and, if the magiſtrates are not unanimous, in order that friendſhip or party may have no influence, the queſtion ſhould be determined by ballot.*

CHOICE OF COMMISSIONERS.

Three Commiſſioners, who are men of judgment, integrity, experienced, and diſintereſted, will tranſact buſineſs much better than a larger number. The act of any two of them ſhould be binding. It is reaſonable that the parties ſhould have a good opinion of the perſons in whom ſo conſiderable a truſt is veſted. The Biſhop of the dioceſe, the patron, incumbent, or other tithe-owners, or the major part of them, ſhould therefore name one Commiſſioner, and a majority of the land-owners chooſe

* If it ſhall appear that a larger proportion of conſents ſhould be obtained, either on the part of the tithe-owners, or that of the land-owners, the author does not object to it. Convinced as he is that the great expence of paſſing a bill through parliament for each pariſh *ſeparately*, and of procuring evidence to attend from diſtant parts of the kingdom, would in many caſes fruſtrate the benefits propoſed; he wiſhes for the conſents to be ſuch as would inſure its ſucceſs if brought before parliament. He does not wiſh for the magiſtrates to have, what ſome may think, too great a power; but that the parties ſhould be enabled to proceed in the leaſt expenſive manner, purſuant to a general act, in the framing of which every proper precaution ſhould be taken.

another,

another, whose names should be produced at the quarter-sessions, together with those of the consenting parties. The magistrates there assembled should choose the third commissioner by ballot. Partiality or friendship may, perhaps, have some influence on the choice of the *parties*. It is presumed the *magistrates* would be particularly careful to choose a man of character, judgment, and experience. But no magistrate should ballot for such third commissioner, or upon the previous question, if he is interested in lands or tithes in the said parish. A commission or instrument in writing, should then be signed and sealed by the justices there present, empowering the commissioners to proceed upon the business, conformable to the general act for that purpose: which commission or instrument in writing should be deposited with the Clerk of the peace, and an attested copy of the same be delivered to the said commissioners.

THE DUTY AND POWERS OF THE COMMISSIONERS.

In the first place the commissioners should give ten days notice at least, in some country newspaper which circulates in the neighbourhood, of the time and place of their first meeting, and should also give like notice in the parish church immediately after divine service, on two Sundays previous

previous to ſuch meeting; and require that all perſons poſſeſſed of, intereſted in, or claiming any tithes, or moduſſes, lands titheable or exempt, do attend, and give in a particular account of the ſame. When the commiſſioners and parties are met, before they proceed to buſineſs, each of the commiſſioners ſhould take and ſubſcribe an oath to the following effect:

> I *A. B.* do ſwear, that I will faithfully, impartially, and honeſtly, according to the beſt of my ſkill and judgment, execute the powers and truſts repoſed in me as a commiſſioner for aſcertaining and ſettling an equitable money payment in lieu of tithes within the pariſh of ———.
>
> So help me GOD!

If any modus is ſet up or claimed by one party, and denied by another, the commiſſioners ſhould in this, as in all caſes brought before them, be empowered to examine witneſſes upon oath. But as the legality of a modus may be too nice a queſtion for them to determine, the attorney of each party ſhould, if required, attend;—and if the matter in diſpute cannot be ſettled, the caſe ſhould be drawn up, agreeable to the evidence, and ſigned by the commiſſioners. The parties ſhould be required to fix on ſome eminent counſel to determine the ſame. If they neglect to do ſo, the commiſſioners may requeſt

request the Lord Chief Baron of the Exchequer, or the Judges at their next assizes, to name one. In default of compliance, the commissioners to submit it to such counsel as they think proper; whose determination should be final.

An accurate survey and plan of the lands will now be wanted. If such is not already taken, a surveyor should be employed for that purpose. A plan of the property will be of great use to the commissioners in the conduct of the business, and should be inrolled with their award. The boundaries and names of lands, as well as the owners, are frequently changed. In process of time, fences are grubbed up and destroyed; several inclosures made into one, or one divided into several; and the ancient names forgotten. Hence it is not uncommon to find that an old terrier, without a plan, is unintelligible. When the commissioners have finished their valuation, and calculated the respective sums to be paid out of the several estates, in lieu of tithes, I would recommend to them to call the several proprietors and tithe-owners together; read over the quality and prices of the lands, and explain, to such as are desirous of information, what principles they have proceeded on. It is a matter of considerable trust and consequence; nor is it necessary that there should be any mystery in the proceeding. Every

man,

man, whose property is at stake, has an undoubted right to give his opinion, and to produce evidence relative to such matters as he conceives to be wrong. However contrary this may be to general practice, I have frequently on Inclosure acts experienced the good effects of it. Men who have cultivated land for many years, and observed it at very different seasons, will sometimes furnish useful information. And it does not follow, that this open conduct of the commissioners should betray them into any concession that their judgment disapproves. The sums which the several estates are respectively to pay in lieu of tithes being determined on, a schedule of the same should be affixed against the door of the parish church where the tithes are commuted for. If no objection is made, nor any appeal intended, the commissioners' award should be drawn up, with a plan and terrier, setting forth the lands chargeable and exempt, the money payments now fixed, and the ancient modusses allowed: these, together with their commission, and the oath they have taken, should be inrolled with the clerk of the peace. A copy of the same should also be lodged in a box or chest, within the church or chapel of the parish where the lands and tithes are situate. If any of the estates should afterwards be divided, and alienated in separate parcels, the plan will always shew the

lands

lands originally charged. And the quota for each parcel, after ſuch diviſion, may be fixed by agreement of the parties, at the time of ſuch alienation, or by two aſſeſſors, with as little difficulty as its proportion of land-tax.

APPEAL.

If it ſhall be thought proper, any of the parties who conceive themſelves to be injured may have liberty to appeal at ſome general Quarter-Seſſions, within four months after the cauſe of complaint has ariſen; giving the commiſſioners one month's notice of the ſame. Though it muſt be confeſſed, that appeals in ſuch caſes are ſeldom attended with any good effects. For it is ſcarcely to be ſuppoſed, that any ſtronger evidence can be produced than that of three diſintereſted and experienced men, whoſe judgment and integrity have recommended them; and who have, with great attention, unanimouſly determined the matter upon oath. But if any one of the commiſſioners ſhould proteſt againſt the proceedings of the other two, an appeal may lie with great propriety.

EXPENCES.

Although it is preſumed that the propoſed commutation will be deſireable both to the clergy and the laity, yet as the inconveniency of tithes is very great

great to the latter, and as the intereſt of the former is but temporary, it is propoſed that the clergy ſhould be exempt from all expence, except a ſhare of that which may ariſe on determining whether a modus is or is not legal.

The general pay of commiſſioners of incloſures, is a guinea per day for time, and half-a-guinea for expences. Whatever charges are incurred on the buſineſs ſhould be borne by the owners of the land, in ſuch equitable ſhares and proportions as the commiſſioners ſhall direct. A power ſhould be given, as in incloſure bills, for tenants in tail, for life, or for long terms, to borrow money, and charge it on the lands, keeping down the intereſt of the ſame. The commiſſioners to direct the application of all ſuch money, and to account with the proprietors, when called on for that purpoſe.

FOR THE SECURITY OF THE CLERGY.

On the preſent eſtabliſhment, the clergyman has a right to his tithes as ſoon as ſevered and divided into proportionate ſhares. He is in no danger from the failure of any tenants, except thoſe of his own chooſing. It ſhould therefore be provided, that whenever a landlord ſhall ſeize for rent, the tenant's effects ſhould be anſwerable for one year's compoſition to the tithe-owner; who ſhould alſo, equally

with

with the land-owner, be entitled to his remedy by diſtreſs. But as this remedy is ſuch a one as every man of feeling, and particularly a clergyman, would wiſh to avoid, it would be proper for him to have the privilege of calling on his pariſhioners to nominate, at a veſtry, two collectors, for whom they ſhould be reſponſible. The clergyman to allow them three-pence in the pound for collecting. If the collectors of the land-tax were to be appointed for that purpoſe, it would be but little additional trouble, and would make it well worth their attention.

METHOD OF VARYING THE PAYMENT.

As the value of money or of the neceſſaries of life riſe or fall, on an average, ſo ſhould the payment for tithes riſe or fall in like proportion. The method of doing this ſhould be as ſimple, and certain, as little complicated, as general, and liable to the leaſt trouble poſſible. If we attempt to regiſter the various articles out of which tithes iſſue;—if different commiſſioners fix theſe values for different pariſhes;—if each pariſh is to be ſeparately conſidered, and regulated at the end of a certain number of years from its firſt payment;—if the fluctuating values are from time to time to be determined by particular markets, or by the prices on particular days, it will lead poſterity into a labyrinth of trouble;

it would be liable to great abuſe, and be productive of much diſcontent and error. In fixing upon one ſingle ſtandard, by which to eſtimate the future value of tithes, in proportion to the value of money, or that of the land out of which they iſſue, perhaps there is no one more proper than that of a buſhel, or a quarter of wheat. Wheat is not quickly of a periſhable nature. Bread is emphatically called the ſtaff of life. And it generally happens, that when it is dear or cheap, other proviſions are dear or cheap nearly in the ſame proportion. Farmers remark this in an ancient proverb—" Down corn, down horn;"—meaning, that the price of horned cattle, or butchers' meat, generally follows that of bread. Some new productions in huſbandry will, often repeated, tire the ground and degenerate. We owe it probably to the goodneſs of Providence, that the ſoil once proper for wheat will, at regular periods, with manure and culture, admit of a repetition with the greateſt probability of ſucceſs. Add to this, that its exportation in years of plenty, and importation in years of ſcarcity, contribute to keep the price of wheat more nearly on a level than that of moſt other articles.

Suppoſe the clerk of every principal market throughout each county was, once a year, at the court of general Quarter-Seſſions, held the firſt

week

week after the Epiphany, to give in upon oath the average price of a bushel or quarter of wheat, on each market-day, in the several months of October, November, and December. In these months farmers generally thresh out great part of their grain, particularly wheat: for as to oats and barley, they generally contrive to thresh, so as for the straw to be given fresh to their cattle, when the grass is gone, and they live in the farm-yard. The sessions being so soon after, the jury, or any persons who attend the court, will have an opportunity of observing, whilst the market-prices are fresh in their memory, whether the average is a fair one. The average price of each market-day in those months being known, the whole may be added together, and being divided by the number of market-days, will give the average price of wheat the preceding year, in each market respectively.

Suppose the markets for regulating the prices of wheat, in Wiltshire, are, Salisbury, Marlborough, Devizes, and Warminster; and that the average price of wheat found as aforesaid is,

	£.	s.	d.		£.	s.	d.
At Salisbury	0	5	2	At Devizes -	0	5	4
At Marlborough	0	5	0	At Warminster	0	5	6
					1	1	0

These added together make one guinea, which divided by 4, (the number of markets) gives 5s. 3d. which is to be entered with the Clerk of the Peace as the average price of a bushel of wheat in the year ——, for the county of Wilts.*

If an act of parliament for this purpose should pass, in or before the year 1790, the average price of a bushel of wheat for each year being proved, and registered as aforesaid, until the year 1800; the whole should then be added together; and an average for that period be taken, and published in the court of Quarter-Sessions, and also in such country news-papers as circulate through that county. In the years 1810, 1820, 1830, and at a like decennial period ever after, the averages should be collected. Those parishes where a commutation takes place before the year 1800, may be varied in the year 1810; and those commuted for from the year 1800 to 1810, may be varied in the year 1820: thus,

* The only reasons for preferring the three months before-mentioned, are, that it will be less trouble, and the prices will be more easily remembered by those who attend the sessions, than the prices for a longer period. The markets at this time of the year are generally full, and the period is sufficiently long to prevent any collusion. Whether the exact average of the year can by these means be obtained, does not seem of so much consequence as, whether the average of the first ten years bears a due proportion to the average collected, in like manner, during a period of any subsequent ten years. However, if six months or the whole year be preferred, for collecting the average of each year respectively, it may be done without much trouble; and the principle will remain the same.

Suppoſe the value of a church living be fixed at 100l. per annum, within the decennial period ending at the year 1800; and it then appears that the average price of wheat collected as aforeſaid, is for that period at 5s. a buſhel; the value of the church living is equal to 400 buſhels, or 50 quarters of wheat. In the next decennial period ending at 1810, we will ſuppoſe the average price of wheat to be 5s. 6d. a buſhel: 400 buſhels at 5s. 6d. is 110l. per annum; at which ſum the value of the tithes ſhould be fixed for the period commencing at the year 1810, and ending 1820: and ſo proceed in the ſame order. The whole ſum which each pariſh or tithing ſhall be advanced or lowered, will thus be regularly obtained at a ſtated period without trouble; and the proportionate ſhare of every eſtate can eaſily be calculated by any ſchool-boy.

It will probably appear to ſome, that it would be better to procure or fix a ſtandard or average value for a period of ten years *previous* to the commiſſioners' valuation in the ſeveral pariſhes reſpectively. This I have conſidered with due attention. Arguments may be adduced in its favour: but upon the whole it will probably not be found preferable to the method before propoſed.

CONCLUSION.

THE laudable society to which this is addressed, have doubtless considered the numerous inconveniencies attending tithes in kind; and I am not without hopes that the hints herein contained will point to a remedy equally desireable to the clergy and the laity.

The clergyman will no longer depend on a troublesome and precarious subsistence, productive of perpetual discord between him and his parishioners. He will know the exact value of his living before he accepts it. His just dues will be secured to him without trouble, and without risk; and he will no longer be charged with ingratitude to his patron, or oppression to his parishioners. The industrious husbandman now secure in the fruits of his labour, a more vigorous cultivation will ensue:—the clergyman and his parishioners may thus be united in one bond of social union, and every disgraceful animosity be forgotten.

Imperfections are sometimes found in subjects less difficult and complex, even where the united wisdom of the legislature has been exerted. The avocations of the writer of this essay have prevented him from extending his observations, or being

being more regular in his remarks. Errors, he fears, might have escaped him, but not through negligence. If he has any claim to merit, it is that of having done his best upon a subject, which has hitherto ineffectually engaged the attention of men of great rank and abilities; and which is, confessedly, of importance to a very considerable part of the community.

B. PRYCE.

Salisbury, Sept. 20, 1786.

Article III.

Strictures on the Husbandry of Turnips, or an Attempt to promote a successful Culture of that useful Root with more certainty *than hath been* generally *practised.*

BY JOSEPH WIMPEY.

[In a Letter addressed to the Secretary.]

Sir,

IN a collection of miscellaneous papers, written by many different persons, it is to be expected, that we find not only a variety of opinions, but often such as are incompatible and irreconcileable. To a reader who wishes to be informed rather than amused,

amused, this is an unpleasant circumstance, and naturally begets embarrassment, diffidence, and distrust. If, by garbling the papers, we could separate *truth* from *error*, *right* from *wrong*, confirm and establish the former, and discountenance the latter, at least as being doubtful, if not groundless, it would well reward our pains.

I chose not turnips for my subject, on account of its importance altogether, but also because the gentlemen of the Committee did me the honour of asking my opinion of a letter respecting the subject.

Turnips have been generally considered as an article of *precarious* culture; but this is not to be taken in an absolute sense; for every thing was made perfect in its kind; and there are few things that vegetate more freely or more certainly than the turnip in its proper season: but, like all other vegetables, it is more or less precarious, as the circumstances attending its culture are more or less favourable.

Nature has set and appointed seasons for her several operations. The spring months are the proper time for vegetation and the growth of plants; the summer months, for consolidating and maturing their

their growth; and the autumn, for reaping, gathering and harvesting the same. The farmer, however, in the cultivation of turnips, is obliged to depart from those established laws of nature, to accommodate the crop to his own convenience. The great use of turnips as food for sheep and cattle, is to supply the deficiency of grass and herbage at a season of the year when little of these are to be got; and that turnips may be in perfection at a time they become most useful, the farmer is obliged to postpone sowing them at least three months beyond the time that would be most seasonable, that is, most favourable to their vegetation. For instance, were turnips to be sown in March or April, as the season might prove most favourable, there would be, I conceive, as great a certainty of a crop, as of any other vegetable usually sown in those months: but the farmer, for the reason before given, being obliged to defer it till the hottest season of the year comes on, his success becomes exceedingly precarious, unless he is so fortunate as to have a few rainy days, or cloudy weather and frequent showers, soon after the seed is in the ground. This I conceive is the true and only reason why the turnip is a more uncertain article than those which are sown in due season.

If theſe obſervations are juſt, the provident farmer will embrace every favourable opportunity that offers for ſowing his ſeed. He fortunately is not confined to a few days or even weeks. He has from the end of May to the beginning of Auguſt, to perform this work, and he had better defer it even to the laſt, rather than ſow when the weather is hot and dry; for in that caſe he may ſow again and again, and loſe both ſeed and labour. But ſhould the weather be ever ſo favourable, that alone will not inſure him ſucceſs: there are ſeveral other things that are equally neceſſary.

1ſt. It is abſolutely neceſſary that the land be very well pulverized. The number of ploughings and harrowings for this purpoſe cannot be aſcertained; that muſt ever depend upon the nature and condition of the ſoil. Twice in ſome land would be more effectual than four times in other; but be the labour whatever it may, it muſt not be omitted.

2dly. It is equally neceſſary that the ſoil be either naturally rich and good, or made ſo by a proper quantity of manure. Turnips never arrive to a good and profitable ſize in poor land, without good manure to promote their growth and puſh them forward.

3dly. It is of great consequence to have seed that is both good in quality and of a good species. I prefer the large green-topped, as being the sweetest and most juicy. Some prefer the red or purple-topped as being hardier; but of which ever sort you sow, if the seed be from the largest and finest transplanted turnips, it is greatly to be preferred, even if it cost double or treble the price of the common sort. I have frequently bought of the seedsmen in London, but it is generally of a mixed kind, and often a great part not worth cultivating. I would therefore recommend it to the farmer to buy the best species he can get, let the price be what it may.*

4thly. As to the quantity of seed, I am pretty much of opinion with another of your correspondents, who advises to be sure to allow seed *enough*, and to that end thinks the safest way is to allow two pounds to an acre, though it is common with many to sow but one. Supposing the seed to be good and the season favourable, a few ounces would be sufficient to stock the land; but as the article is so very precarious, it is by far the safest way to allow seed in plenty, and reduce them afterwards by well harrowing the ground.

* This remark of Mr. Wimpey's is of great consequence: and for the reasons he assigns, the Secretary of the Bath Society makes a particular point of keeping a supply of the finest turnip-seed for such gentlemen and farmers as apply to him.

Lastly.

Lastly. The greater your success in providing a good plant, the greater will be the necessity that the crop be well and carefully hoed; without this, the great advantage to be derived from a good crop of turnips, would in a great measure be lost. Twice hoeing is often sufficient for this purpose, especially if the land be pretty clean; but if it be foul, three times is hardly enough. Hoeing in many places is not well understood, although it be an operation of very little difficulty. Practice is necessary to give dexterity to every kind of work: but a labourer, who has been used to work in a garden, and knows the use of a hoe, would not only perform it well himself with a few hours' instruction, but could teach all the labourers in a parish in a few days, which would greatly reduce the price of that business, it having been exorbitant hitherto in many places.

The business, however, might be made easy, and much expedited by well hoeing the turnips as soon as they arrived at a proper stage of their growth; that is, when they have four leaves; and where the turnips are thick, they might be well harrowed a second time, at the distance of a fortnight or three weeks. This would not only thin the crop, but also greatly improve and encourage the growth of the remainder. In this situation the hoers would readily diftinguifh all that were proper to be cut up from thofe that are to remain, whereas, fhould it be deferred till they are over-run with charlock and other noxious weeds, the labour and difficulty would be more than doubled, and could never be performed fo well. I have feen a field of turnips fo entirely over-run with weeds, that the hoer worked as it were in the dark, and chopped away at random. Three weeks or a month fooner, the work might have been done at half the expence, and to more than three times the advantage.

As to the mode of planting, I am of opinion that the broadcaft is the moft productive, if the hoeing be properly performed and in good time. However I am much inclined to think, that the mode of fowing turnips between beans planted in rows, as recommended by feveral of your correfpondents, is a much more certain means of infuring a crop. It exactly correfponds with all my obfervations on the fuccefsful vegetation of that root. A confiderable degree of moifture is neceffary to the rapid vegetation of that very juicy root, and nothing retains moifture equal to fhade; and fhade can be obtained and fecured by no means fo effectually on a large fcale, as in the intervals of tall growing plants, as beans or wheat planted in drills.

My

My experimental field, of about ſeven acres, is now drilled with wheat on three-bout ridges, about four feet and a half wide. It was horſe-hoed in December, and I intend ſhall be horſe-hoed again in the ſpring and ſummer, as the ſeaſons arrive; in that caſe the mould in the intervals will be in very fine tilth for turnips, with which I intend to ſow them. Theſe may be hand-hoed whenever it becomes neceſſary, notwithſtanding the wheat; and as ſoon as that is harveſted, the ridges it ſtood on may be ploughed, and the turnips horſe-hoed, and perhaps repeated before winter. The crop I propoſe ſhall be fed off in January and February, time enough to plant the intervals on which they grew with beans the beginning of March; horſehoing the intervals, as the growth of the beans will permit, tu prepare them for potatoes to be planted between the beans the latter end of April or beginning of May.

If this method ſhould be attended with the ſucceſs I expect, the land may be continually planted with a double crop, that is to ſay, with wheat and turnips one year, and with beans and potatoes another, in alternate ſucceſſion. If this courſe of cropping ſhould be found to exhauſt the land, more than the horſe-hoeing could repleniſh, which I do not think very probable, a moderate dreſſing of dung

dung might be given every fourth year as ſoon as the turnips are off, to prepare the land for beans and potatoes; the extra expence of which, ſhould it be found neceſſary, would probably be amply repaid by the increaſe of quantity. Indeed the benefit would not terminate here; for as one of your correſpondents has, I think, rightly obſerved, it is far better to manure for turnips the preceding year, than immediately before ſowing them; and I am ſure it is for wheat, eſpecially if the manure be not thoroughly digeſted and become inoffenſive.

Whether plants from new or old ſeed are moſt ſecure from the depredations of the fly, is, perhaps, a queſtion, which cannot be eaſily determined even by experiments; for concomitant circumſtances are frequently ſo much more operative and powerful as to render the difference between them, if there be any, imperceptible.

It is, however, in the knowledge of every practical man, that new ſeed ſprouts or vegetates ſeveral days before old, and I think more vigorouſly; and it is equally well known, that the healthy and vigorous plants eſcape the fly, when the ſtinted and ſickly ſeldom or never eſcape them. It ſhould ſeem then, that new ſeed, *ceteris paribus*, is more ſecure from the fly than old, and for my own uſe I would always prefer it.

That

That old feeds are preferred to new in fome articles by experienced gardeners is very true, and I believe with good reafon; but this furnifhes a reafon againft giving a preference to old turnip feed, contrary to what it is brought for. Old melon and cucumber feed is preferred to new, becaufe the plants from old feed are far lefs luxuriant and more fruitful. In a former paper we obferved, that luxuriance and fructification are very different things; and in a few, perhaps in no genus of plants, are they ftrictly compatible; but the roots of the turnip can never be too luxuriant, and the more they are fo, the more fecure they are from the ravages of the fly.

Many are the *noftrums* for the prevention or remedying the evils of this deftructive infect; but like a charm for the cure of the ague and the tooth-ache, they are found to be equally fabulous and quackifh. It is certainly very bad reafoning to conclude, that becaufe certain things are difagreeable to our fmell and tafte, they muft neceffarily be fo to creatures of a different kind;—and yet from this fource their recommendation feems to originate. From the great encomiums beftowed on *elder*, I was in great hopes a fpecifick remedy had been found; I therefore determined to give it a fair trial:—accordingly I repeatedly drew elder branches, not only over beds of of young turnips, but a variety of other plants; I whipt the ground with them, and ſtrewed the leaves, tops, and tender ſhoots over the beds; and finding all this totally ineffectual, at length I made a very ſtrong decoction in boiling water, and, when it was cold, watered the plants with it ſeveral times. All this had juſt as much effect, and no more, as walking round the beds in the ſuperſtitious garb of a magician, and chaunting *Abracadabra* at every turn.

I am quite of opinion that nothing has yet been diſcovered which is at all adequate to the purpoſe, further than it may invigorate and promote the growth of the plants. To this end aſhes, ſoot, or a rich compoſt of lime and dung, if uſed in ſufficient quantities, may be deemed ſpecifick; but the beſt means of uſing them is, either to ſow them with the ſeed, or rather by themſelves immediately before, and to well harrow them in, that they may be completely incorporated with the ſoil. This for the moſt part would ſo much invigorate and encourage the growth of the plants, as to be an overmatch for the moſt vigorous attack of the fly.

If I might be indulged in a wiſh, I would make it a moſt earneſt one, that no writer in future would advance any thing for a fact, which he himſelf hath not had full experience of the truth of. Nothing can be more inimical to the laudable intentions of the ſocieties eſtabliſhed for the promotion of uſeful knowledge, nor can any thing reflect more diſcredit on their earneſt endeavours to promulgate the ſame, for the general information and benefit of mankind, than promiſcuouſly blending fable with truth, and giving chimeras for diſcoveries, which never exiſted but in the imagination of the writer. The *elder* noſtrum above-mentioned has, I believe, diſgraced almoſt every repoſitory of papers on theſe ſubjects, which hath been publiſhed for many years.

To this I would add another wiſh, which is, that no writer in future would communicate any thing to the ſociety but original papers, without quoting the author from whom ſuch writing was copied or extracted. This would enable the Committee to judge of the propriety of publiſhing the ſame, and often prevent the very uncandid impoſition, which is too often practiſed, of paſſing extracts for originals, and abſurdly and diſhoneſtly cauſing the ſame thing to be publiſhed many times, much to the diſappointment and loſs of the purchaſer.

One offence of this kind I can point out in the 2d volume of your letters and papers, from a gentleman

tleman in Devonshire, signed C. H. in Article XXVIII. on the nature and effects of lime as a manure, which he gives as the united effects of his own experiments and observations on the subject; whereas the whole was extracted from a book entitled "RURAL IMPROVEMENTS," published by myself some years before, as will clearly appear to any one who will take the trouble of comparing the chapter on Lime, page 201, in my book, with the said article.

He is not the only writer who hath purloined from my book; for half a dozen, at least, which have come to my knowledge, have played the same nefarious game. One author (Mr. FORBES) has copied about thirty pages from the said book, but then he has very honestly told the reader from whence they were taken. There is something so very mean, uncandid, and disingenuous in plagiarism, that it is much to be wished an indelible stigma were to be fixed on every offender, to discountenance and prevent a practice so very dishonourable.

NORTHBOCKHAMPTON,

Feb. 8, 1787.

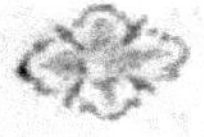

Article IV.

Of the Necessity of adapting or suiting the Crop to the Nature, Condition, and Circumstances of the Land to be planted; with an Account of an Experiment to ascertain the Quantity of Butter and Cheese producible from a given Quantity of Milk.

[By Joseph Wimpey.]

THE first and principal object of every husbandman is, to obtain the most profitable crops from the land he occupies. In order to this, it is absolutely necessary, that he suit the articles he plants to the nature of the soil. When art is made to co-operate with nature, our hope is founded on rational grounds. To act in repugnance thereto, is to sail against wind and tide, and there can be little or no hope of making a successful voyage. It is true, that amazing things may be effected by unremitting perseverance and unlimited expence: but the husbandman's province is not to enquire after what is possible, but what is profitable; not what may amuse the curious, but what will reward the diligent, for the benefit of the community of which he is a member, as well as for his own.

Though

Though the lands of these kingdoms consist of a very great variety, they may fitly enough be ranged under two general heads.

First, all such as naturally produce little of any value, either for man or beast; but require the art, labour and attention of the husbandman, to bring them into a state of cultivation, so as to render them useful and profitable. Here the plough becomes a necessary instrument in breaking up, dividing, and pulverising the soil; and hence such lands are denominated—*arable.*

Under the second head may be placed, all those lands which spontaneously produce grass and herbage proper for the feeding and fattening of cattle, the production of milk, of butter, and of cheese. The occupier of the former is properly a corn-farmer; of the latter, either a grazier or a dairyman; and it sometimes happens, that all three functions are exercised in some degree by the same man, as may best suit the different parts of his farm.

From hence it follows, that occupiers of land, who pursue their business upon principles of reason and œconomy, have no choice, whether they will be corn-farmers, graziers, or dairymen. The occupiers of the former are *necessarily* corn-farmers;

 for

for without the plough and its auxiliary inſtruments, their fields would ſoon recur to their original wild ſtate, and be overrun with furze, briars, and brambles, and ſuch unprofitable productions as would be of no uſe or value. The occupier of the latter, is as much bound by his intereſt to appropriate his lands to feeding or to the pail, as the other is by neceſſity to plant corn; for when nature has given herbage proper for ſuch purpoſes, the land is of much more value, and its profits to be acquired with far leſs labour and riſque, than from arable land.

It is true, indeed, that ſome have been ſo infatuated, as to plough up good meadow and paſture land, and relinquiſh a good and certain gain upon a viſionary and abſurd expectation. And to ſuch, and ſuch only, can the reproof of the ingenious writer of Article IX. in the Third Volume of your papers be applied. All ſuch are juſtly reprehenſible: but ſurely the occupiers of arable land, as ſuch, are by no means cenſurable. To expect they ſhould make butter and cheeſe from land to which nature has denied graſs and herbage, would be a taſk of more than Egyptian difficulty. As it is neceſſary to have butter and cheeſe to eat with our bread, it is full as much ſo to have bread to eat with our butter and cheeſe; and the moſt certain way of

obtaining

obtaining plenty of each is, to appropriate the land to the production of ſuch crops as are moſt ſuitable to its nature.

It is admitted, that ſome land has been very imprudently broken up and converted to arable, which was of much more value to the occupier, while it was in graſs; at the ſame time it cannot be denied, that ſome are as injuriouſly prejudiced againſt the plough, and will not conſent to have unfertile paſtures broken up, although they might be greatly improved thereby. I have now in my eye ſeveral fields not far from Waminſter, which would be worth double their preſent value to the occupier, if the owner would conſent to their being ploughed and planted with corn at proper intervals.

Paſture fields, when become hide-bound and moſſy, bearing little elſe but a fine wiry graſs, almoſt as void of ſubſtance as ſuſtenance, might be expeditiouſly and effectually improved by the plough. Were ſuch lands planted with wheat on the ſod, as practiſed in the county of Norfolk, and the winter after well manured, and planted with beans the March following in rows with three feet intervals, well horſe-hoed till June, and then ſowed with turnips to be eaten by ſheep the following winter; then in April to be well prepared and

ſowed

ſowed with barley, and graſs-ſeeds ſuited to the ſoil; there cannot be a doubt, but that the produce the three years ſucceeding the crop of barley, would be much more conſiderable than it would have been in the ſix years, had it continued the whole time in its natural ſtate. So that the net produce of the three crops of corn would be ſo much clear gain to the occupiers, and proportionally beneficial to the public.

Upon the whole, I think, it may be fairly concluded, that for the loſs of every ton of herbage that has been ſuſtained by means of the plough, twenty, at leaſt, have been gained by the well-timed uſe of it. Moſt, almoſt the whole, of the improvements made in huſbandry in the courſe of the preſent century, have been by the prudent uſe of the plough. Turnips, clover, *all* the artificial graſſes, eſculent roots, herbs, and plants, ſo far as reſpects field culture and the feeding and fattening of cattle of every kind, have been obtained by its uſe ſolely, as none of them can be cultivated extenſively without it. Therefore, true as it is, that butter and cheeſe, and ſome other articles, have advanced almoſt double their price in the laſt thirty and forty years; and true as it may be, that graziers and dairymen pay their rent more punctually than little corn-farmers, or the occupiers of ſmall arable

farms;

farms; it can by no means be accounted for upon the principles of an undue and imprudent attachment to breaking up meadow and pasture land. What are the proper and genuine causes of these effects, we may endeavour to explain hereafter.

The writer of the article above-mentioned was exceedingly misinformed, respecting the comparative value of cheese made of milk, which had been completely skimm'd, and what had not been skimm'd at all. The difference in price is, at least, four times as much as the sum he mentions. Skimm'd cheese, I have been credibly informed, hath been sold at Warminster fair, within about four years last, for 12s. 6d. per hundred; whereas the best rammill, say raw milk cheese, sold for from 38s. to 42s. per hundred in the same fair, and prime cheese from the best dairies for 46s. or 48s. The medium prices of the three different kinds, that is to say, of skimm'd, of half new and half skimm'd, and of milk not skimm'd at all, have been 15s. 28s. 40s. per hundred.

As to the best course of experiments respecting the comparative value of butter and cheese, Mr. Billingsley, in his very judicious remarks on the said article, has given the true, and therefore the best general answer to the question proposed. For both

both produce and prices are fo varied by local circumftances, that, as he obferves, " no fettled invariable rules for the management of the dairy can with any propriety be eftablifhed." The quantity of produce of each article fpecified, may be eafily afcertained on any dairy, but the fuperior advantage of any courfe can be determined only by the demands which arife or are promoted by peculiarities of fituation.

It is afferted in the faid article, " That a tenant of 60l. per annum, in a dairy farm, will get money, while a corn-farm of the fame fize will ftarve its occupier, (though perhaps the former gives 15s. per acre for his land, and the other but 10s.) is felf-evident." This is by no means a neceffary truth arifing from caufes eftablifhed in the nature of the thing, but has its foundation in artifice, as will be fhewn hereafter. However, this is not the interefting queftion. Is it felf-evident, or by any means demonftrable, that a corn-farm of 10s. per acre, which would ftarve its occupier, is by any method convertible into a dairy-farm; and that if the plough fhould be abandoned, and fuch land fuffered to recur to its original and natural produce, as in that cafe it muft do, would it not ftarve the occupier, even if he gave but 5s. per acre for fuch land?

It ſeems to have been totally forgotten, that the lands of *all* corn-farms, be they little or great, were originally paſture, and in that ſtate applicable *only* to grazing or the dairy: and many hundred thouſand acres of ſuch land, worth only in that ſtate a ſhilling or two an acre, have been improved by means of the plough, to 10s. 15s. 20s. per acre, and ſome much more. Relinquiſh the plough, and thoſe very lands would, by rapid degrees, revert to their original ſtate of unproductiveneſs, and conſequently would be of no more value. It is with lands, as with the occupations of men: ſome are incomparably more lucrative than others; but all men cannot be of thoſe occupations which are the moſt lucrative, nor have they talents for it. Ralph may poſſeſs every qualification neceſſary to conſtitute an excellent ploughman, but, probably, no education upon earth could qualify him for a Prime Miniſter, or a Lord Chancellor. So, many fields by proper culture would bear excellent crops of wheat, to which nature has denied herbage proper to fatten a rabbit. Individuals in certain ſituations may ſuffer by ill-judged converſion of land proper for grazing to arable; but I am of opinion, the practice is neither of ſuch extent or magnitude, as to advance the price of butter and cheeſe, even ſo much as a farthing a pound in the general market. Admitting

then

then that the little corn-farmer of 50l. or 60l. per annum, with great labour and assiduity finds it difficult to live; while the dairy-farmer of the same rent, not only carrieth on his business with incomparably more ease, but is getting money at the same time; also that butter and cheese are advanced at *least* a third of their present price within these 20 or 30 years; I say, admitting these for facts, which I believe cannot be denied, we will proceed as briefly as we can to assign the true and genuine causes of the same.

It is a maxim generally allowed, that unless a farmer makes three rents he cannot live. A dairy-farmer, then, 20 years ago, whose rent was 60l. per annum, by this rule made 180l. per annum; so that having paid his rent, he had 120l. left; labour, expences, loss of cattle, and incidental charges, having been usually reckoned another third, the remaining 60l. was for the maintenance of himself and family. But if the produce of dairy-farms be advanced a full third of its present price, what sold then for 60l. will now yield 90l. and consequently the gross amount, which was 180l. then, will be 270l. now; from which 60l. being deducted for rent leaves 210l. and from that sum another 60l. as before for expences, &c. there remains then 150l. so that upon these principles the dairy-farmer has

a net

a net 90l. per annum, for his maintenance and profit, more than he had 20 years ago. It can be no wonder then, that he punctually pays his rent, and saves money. But it may be said, and indeed very truly, that rents have been generally raised, especially on little farms, nearly in the same proportion, and on some considerably more, and that so much must be deducted from the sum above-mentioned. The remark is just, and the account being rectified accordingly, it will stand thus: instead of 60l. for rent, we must allow 90l. consequently the additional 30l. is to be deducted from 150l. which reduces his net gain to 120l. which is just double what it was 20 years ago, and a very pretty income it is for a man of so small capital, and in so little business. Let us next enquire how, on the same principles, matters stand with the corn-farmer, who is represented as being in a starving and ruinous condition.

The corn-farmer is supposed also to occupy a farm of 60l. per annum; that he, like the former, makes three rents, one for his landlord, one for expences of all kinds, and the other for his maintenance, &c. But his expences will be far more considerable, as well as his labour and care, than the dairy-farmer's, and the surplus of these expences must come out of his share. His farm has been equally raised with the former; therefore he now

pays

pays 90l. instead of 60l. he paid before; the additional 30l. being deducted from 60l. his share, leaves only 30l. to maintain his family, and make good the extra expences of the second share. He has no resource to an advance of price in the produce of his farm like the former, to enable him to pay his advanced rent, which may be easily seen, by comparing the average prices of corn for the last 20 years, with those of the 20 years immediately preceding, which I fear will be found to afford him little assistance. If, then, it was with difficulty enough he made both ends meet before his rent was advanced, how is it possible he should live now upon an income reduced one half, say from 60l. to 30l. or more probably from 40l. to 20.? What is to be done then? To convert a farm that is properly arable, to a dairy-farm, is impracticable; and were it not, should it be generally practised, it would entirely defeat its own purpose. The remedy, and the only remedy, seems to consist in a reduction of the rents of such farms, and the farmers adopting the modern improved culture, recommended by the very intelligent Mr. BILLINGSLEY, of "judiciously blending *arable* and *pasture*," but I think seldom "in the proportion of *three* of the latter to *one* of the former." If the farmer could get two good crops of artificial grasses, to two or three of corn,

corn, which I think would be more ſuitable to moſt lands, perhaps his crops of both would be more beneficial, than on any other diviſion. But every one's mode of practice muſt be governed by the peculiar circumſtances of his farm.

But it may be aſked, if breaking up paſture lands, and converting them to tillage, is not the cauſe of the advance of price of butter and cheeſe, what is?—I anſwer, what would raiſe the price of any commodity whatever; it is foreſtalling, ingroſſing, and monopolizing. And perhaps there is no article in the large circle of commerce, that is ſo much the ſubject of thoſe pernicious arts, as butter and cheeſe.

The cheeſemongers in London, many of them at leaſt, are men of large capitals, who have correſpondents, agents, and factors, in many, I believe in moſt, of the conſiderable dairy counties in England. The prices of butter in large dairies are uſually fixed and agreed on at the beginning of the ſeaſon; and whether the year proves plentiful or otherwiſe, it makes no difference in the price. What is bought dear, will always be ſold dear, where there is no oppoſition or competitor in the market. I was once at Axminſter, when no bread and butter could be had with our tea; the reaſon being aſked, the miſtreſs of the inn aſſured us it

frequently

frequently happened that an ounce of butter was not to be got in town, unleſs on a market-day; for all the great dairies were under contract with the London dealers, for all they make, at a fixed price, which made it both ſcarce and dear. At the time ſhe ſaid this, there were 100 tubs of butter piled up in the gateway of the inn, in readineſs for the London waggons. Upon enquiry I found, the current price was 7s. 6d. per dozen wholeſale, and that the town and neighbourhood was ſupplied by ſuch little dairies only as were thought below the notice of the wholeſale dealers.

A ſimilar mode of practice is followed in regard to cheeſe. The great dealers in London long ſince inſtituted a club, and hold a weekly meeting to regulate their affairs. They employ agents or factors in Cheſhire and Lancaſhire, to buy up the cheeſe made in thoſe counties, which is done by agreement for whole dairies; they have ſeveral ſhips in their employ, which perform almoſt the whole carrying buſineſs between Liverpool and the metropolis. Not one of theſe ſhips is permitted to carry ſo much as a cheeſe for any one but the company. At their weekly meetings, they ſettle the quantity to be brought by each ſhip, which they proportion to the demand, being very careful that the town ſhall not be overſtocked, but kept rather hungry,

as

as all the rest of the dealers are supplied by them, as indeed is almost all England; for a good Cheshire cheese is hardly to be got even in Cheshire, as I have often heard from the masters of those vessels, who are frequently employed to buy Cheshire cheeses in London, and carry them back again to gentlemen in those counties, who can get none that is good at home. From hence it is easy to conceive, how much the price must be enhanced by two commissions, two freights, and the profit of at least one commissioner, but very commonly of two.

It may be said, this respects a county or two only; but it may as truly be said, that a similar practice obtains almost through every dairy county in the kingdom. Jobbers have established themselves almost every where, who either buy all they can immediately from the dairies, or constantly attend the markets and fairs in the neighbourhood of the dairies, and ingross large quantities, which infallibly advances the price of the whole immoderately. For instance, in Wiltshire, the jobbers, 20 miles and upwards round Marlborough, constantly attend that market, where they buy up and contract for very large quantities of cheese for the ensuing fairs; that is to say, for Newbury, Andover, Weyhill, and Reading; from whence, if they are not

 offered

offered a price to their minds, it is ſent by water to London, which is a market that infallibly takes off every thing. But here it is got to the end of its journey, and muſt be ſold for whatever it will yield; and this is the reaſon why not only cheeſe, but all ſorts of grain, &c. are uſually ſold below the average price which generally obtains throughout England; often indeed conſiderably under what they yield in the place where they grew, or were manufactured.

Some 30 years ſince, it was uſual for cheeſe to be ſent immediately from the dairy to the fair in ſurpriſing quantities, and the price was then determined by the proportion the quantity bore to the demand. The bleſſings of propitious ſeaſons were then enjoyed in common, and the conſumer came in for his proportion; but now this natural and regular courſe is almoſt totally perverted by the jobber, and the price is no longer governed by the above proportion, but by the price it coſt the jobber, and the profit he thinks fit to put upon it. He is not obliged to comply with the current price, like the dairy-man, who had no other reſource; the buyer muſt conform to his price, or he will try it again and again at other fairs, till he ſucceeds, or in caſe of a diſappointment, at laſt it is ſent to the never-failing market—LONDON.

Notwithſtanding

Notwithſtanding the prolixity of this paper, I find myſelf under a ſtrong temptation to add a few lines in behalf of the little, and too often much diſtreſſed farmer. If the occupations of men were to be eſtimated by the ſervice they render the publick, and the pitiful recompence ſome have in compariſon with others, I believe it would be extremely difficult to find any claſs of men who are ſo richly entitled to favour and encouragement as the little corn-farmer. His toil and anxiety are inceſſant; his labour, from the riſing of the ſun to the going down of the ſame, and often much longer; his diet the pooreſt; his clothing, lodging, and other accommodations, as mean and comfortleſs as can well be conceived. Theſe are all the recompence he has in general, for his indefatigable and unremitted labour in providing the neceſſaries of life for the reſt of the community; for to the little and middle farmer, are we chiefly indebted for the well-timed ſupplies of thoſe things, without which we could not ſubſiſt. It is the little and middle farmers, who ſupply the market from harveſt to Chriſtmas and onwards. They muſt raiſe money to pay ſervants' wages, tradeſmen's bills, taxes, rent, &c. &c. The "whim or caprice, or ſuppoſed advantages," which are aſcribed to corn-farmers in general, are applicable only to the great and opulent, who are able to monopolize, as well as cultivate: and ſome, through an avaricious ſpirit, would withhold from market, till corn, &c. advanced to a price that would ſtarve the reſt of the people. I therefore repeat it, the little farmer, whether he be a corn or a dairy farmer, provided he prudently appropriates his land to the production of the moſt beneficial crops, cannot receive too much countenance and encouragement from the publick, not only to whoſe convenience, but ſubſiſtence and well-being, he devotes a ſlaviſh and moſt laborious life.

I have already obſerved, that the diſtreſſes of the little corn-farmer are in part owing to the great advance of his rent; but whenever, in the viciſſitude of human affairs, the prices of grain ſhall fall to that low price which a ſeries of fruitful years hath ſometimes produced, as for example, between 1730 and 1750, rents will tumble faſt indeed, but unfortunately the poor farmer muſt tumble firſt. Corn being once at the low price it ſold at then, no device or artifice whatever could keep up rents to the pitch they have attained at preſent. No routine of crops yet propoſed, though made with the greateſt judgment, would enable the little farmer to ſtand his ground, and ſatisfy his landlord.

But

But though ſo large an advance of rent is one, it is not the only reaſon of the little farmer's difficulties. It is generally allowed, that to do well, a farmer's capital muſt at leaſt be equal to three times his rent; but I am well ſatisfied, if it were equal to five rents, it would be vaſtly better both for himſelf and his landlord. What a miſerable chance then muſt both run, when the farmer is ſcarcely poſſeſſed of ſtock equal to a ſingle year's rent, which I am afraid is too frequently the caſe. The true judgment of the landlord conſiſts then in letting his farms to tenants, whoſe capitals are fully adequate to the rent they are to pay, and at ſuch rents as with good management they may be able to pay. Such rents would be *real*, and might be depended upon; but farms let at rents racked up to the higheſt pitch that tenants with little or no capital will conſent to give, are properly not *real*, but *nominal* rents, and ſuch too commonly end in the ruin of both the farmer and his farm. The ſure way, therefore, for a land-owner to have the rent of his corn-farms paid as punctually as his dairy-farms are ſaid to be, is to take care that his tenant's capital be fully adequate, and his rent proportioned as favourably to its improved produce, as the dairy-farmer's is to the natural produce of his.

As experiments to determine the comparative value of butter and cheeſe have been thought of ſome importance, I take the liberty of preſenting one to the ſociety. It is on a ſmall ſcale, but made with great care and exactneſs. One hundred and five gallons and a half of milk were properly diſpoſed in pans for ſkimming off the cream. It produced 36 pounds of butter, and 60 pounds of ſkimm'd cheeſe. The low average of good butter, in this neighbourhood, is 8½d per pound. And the ſkimm'd cheeſe was ſold for 2d. per pound. I am informed this ſort of cheeſe, three or at moſt four years ſince, ſold only for a penny farthing, or at moſt three half-pence per pound.

	£.	s.	d.
36lb. of butter at 8½d. - -	1	5	6
60lb. ſkimm'd cheeſe at 2d. -	0	10	0
Total -	£.1	15	6

Of a like quantity of milk, ſay one hundred and five gallons and half, were made 106lb. of raw-milk cheeſe, and 6lb. of whey and butter. The cheeſe at two months old was worth at leaſt 3½d. per pound, and the whey butter ſold at 7d. per pound.

	£.	s.	d.
106lb. raw-milk cheeſe, at 3½d.	£.1	10	11
6lb. whey butter for 7d. - -	0	3	6
Total -	£.1	14	5

From

From this experiment it appears, that when the butter and cheefe, of each fort above-mentioned, will fell for the above prices, a fmall advantage lies on the fide of butter and fkimm'd cheefe. It amounts to 13d. only in 1l. 15s. 6d. which is about 3 per cent.

Butter from half new milk and half whey would be of a middle quality between the other two, and the coft price of courfe muft be fo too; and fo muft cheefe from half-fkimm'd milk; but each of thefe may be varied in goodnefs according to the proportion of milk and whey, and of the milk fkimm'd and unfkimm'd; fo that the price of the latter may be varied from 20s. to 30s. per hundred. But the advantage of following either courfe depends upon local circumftances, as was obferved before; and the dairy-man, to acquire the greateft profit, muft regulate his mode of practice according thereto.

North-Bockhampton,
Dec. 6, 1786.

ARTICLE V.

Practical Observations on sundry Letters in the Third Volume, and on the Advantage of Friendly Societies.

[In a Letter to the SECRETARY.]

SIR, *Stifted-Hall*, *Essex*, *June* 20, 1787.

I Beg leave to acknowledge the favour of the 3d volume of the Papers of the Bath Society. At the 53d Article, Mr. LAMPORT, from an observation of a common husbandman, enquires the reason, why *old* turnip-seed should escape the ravages of the fly more than the new?—I take the liberty of informing you, that on the supposition of its doing so, our best farmers mix together for sowing half of each, in the whole a quart for an acre broadcast; and find that the *new* seed vegetating first, and probably possessing some greater sweetness, affords occasionally sufficient food for the fly, till the whole is grown strong enough to escape. The mere chance of this benefit, for a vegetable crop so very important, is a sufficient inducement for its practice; they are here never hand-hoed *less* than twice at 8s. an acre, beer included.

Sir THOMAS BEEVOR's account of his trial, in the *same mode of culture*, of the turnip-rooted cabbage, is of of very great importance, and ought to be followed by every winter-grazing farmer: ſince a *ſmall* portion of this hardy and abiding plant, cultivated *exactly* as the common turnip, would remedy the great inconvenience and expence that is commonly ſuffered in the beginning of ſpring, when on all ſoils, more eſpecially the heavier ones, turnips muſt be gone; and no material graſs-feed can in common, or to any great degree, be had. From a few plants of this turnip-rooted cabbage, which I raiſed when it was firſt talked of, it ſeemed more ſuitable to our ſtronger ſoils than the common turnip, and far more capable of bearing froſt; when boiled, its root has much of the cabbage flavour. An acre or two of this, *as a ſure reſource*, even if a fallow followed it, would be valuable; but to an Engliſh farmer, beſides buck-wheat, there are ſo many ſeeds, roots, and graſſes, for ſummer-ſowing, ſo well known, that the fallow will probably be unneceſſary.

In a note on your 43d Article, it ſeems doubted whether four horſes be equal to the due cultivation of one hundred acres of arable; but it ſhould be recollected, that *with us*, no inconſiderable portion of this is in rotation after corn, under artificial paſturage; and it is *this* plan of modern farming that is the uncontradictable fact in ſupport of all incloſures, as it nearly inſures an equal quantity of

every

every ſort of cattle to be kept, and an equal quantity of every ſort of corn to be raiſed, on half the land, as was done before the incloſure on the whole; whether it was then, for the former inſtance, open grazing common, or for the latter, open arable field. How population can be injured, or rather, how it ſhould not *thus* be promoted, may be diſtinguiſhed by the jaundiced eye of ſome ſpeculative politician, but is not at all perceivable by any one of common ſenſe and experience, who can laugh at theory, (and happily he has, to conſole him in his want of knowledge, plentiful food for laughter) when he finds it totally irreconcileable with reality.

I take the liberty of confirming my obſervations in your 38th Article, on the Cow-Graſs Ley [*Trifolium Pratenſe.*] I have the fineſt plant of full-eared wheat in this neighbourhood; my thick-ſown rye-graſs was fed (even in this ſpring) in the middle of April, a benefit which I endeavour to inſure, by always affording that field under the corn, of which it is annually ſo thickly ſown, a coat of manure ſoon after the corn is harveſted. My acre of carrots, which in the ſame article I mentioned as being from laſt year's drought, together with my not affording them, though ſown on a wheat ſtubble, (ſo the trial was ſufficiently indelicate) but one

ploughing,

ploughing, produced (including a ſmall patch, which I tried advantageouſly with parſnips) two hundred and thirteen buſhels; the greater part of both were taken up in March, and given to my horſes. The turnips, which I harrowed in over the carrots, were for the year (a failing one) a ſufficient plant for the wants of the dairy, after about two acres of cabbages and borecole had been conſumed; which turning out, like the carrots, a very imperfect plant, the vacancies were filled up by *every ſpecies of refuſe plants* of the cabbage kind that were uſeleſs in the kitchen-garden, and our œconomy was rewarded, by not only ſaving this various herbage for the cattle, but by having our own table unexpectedly treated, from this whimſical field mixture, with plenty of very forward and fine green brocoli.

Such ſlovenly farming did not, it may be ſaid, merit ſuch plenty; but it ſuggeſted to us, that, with more becoming neatneſs, a field of about three acres would, for the eſtabliſhment of any country gentleman, be more advantageouſly cultivated in this than in any other mode of agriculture; ſince by this gardening at the cheapeſt expence, under the plough, all the common winter and ſpring herbage, potatoes, carrots, parſnips, turnips, cabbages, borecole, and brocoli, might be annually

raiſed

raiſed in ſufficient quantities for the houſe, and for all the cattle uſually belonging to a little farm. By changing theſe different ſpecies, the one ſucceſſively after the other, into different parts of this *kitchen-field*, and keeping it neatly hoed, it might, as any other garden, always be cropped under this very profitable as well as moſt comfortable culture.

In order to contraſt my conduct of laſt ſpring, I had in the preſent (after coating it with a little manure) half an acre *dug*, and ſown with ſix pounds of carrot-ſeed; the digging coſt 1l. the ſeed 5s. and three hand-hoeings juſt compleated 1l. The plant is one of the moſt exact and promiſing that can be ſeen.

A neighbour of mine (who on one ploughing of a graſs ley raiſed laſt year from ſix to ſeven hundred buſhels of carrots per acre, and very profitably fattened oxen with them) took up in October 1400 buſhels, and after topping and drying them a little in the field, flung them promiſcuouſly into an out-houſe with a ſlight covering of ſtraw, where they remained for occaſional uſe, if ſnow or froſt prevented the gathering thoſe in the field, or as the reſerve till thoſe were conſumed; which was the caſe, not being uſed till March, when they were ſo firmly ſound as to appear probably more nutritive than

than any taken then fresh from the field. I never saw, as far as I could conjecture, a more advantageous piece of culture; nor where the land (as in the two preceding years, from the crop of barley after the carrots they had experience of) seemed from a vegetable crop in such a promising state as this. But I must observe on carrots and potatoes, it is not the crop produced from a fresh soil, but that, where like turnips they have been cultivated in a regular rotation after corn, and for a series of years, which must determine their fair value and use, both for consumption, and preparatory to whatever corn crop may succeed them.

Turnips have had a long, and cabbages some trial, and, with carrots and potatoes, seem to promise a vegetable crop after a corn one, suitable, either one or the other, to almost all the various arable soils in this kingdom.

In Article 19th, on Mr. ANDERDON's drill-culture of beans and turnips, you justly observe the same soil cannot be suitable to two crops of such an opposite nature. But the farmer, in the rotation of his crops, under the common husbandry, and from the necessities of his stock of cattle, must frequently hazard vegetable ones on soils little congenial to them, and rest his chance of success on that of the

seasons;

ſeaſons; upon the whole, he probably, if cautious, gains an advantage. It is on this plea the greater part of turnips are cultivated here; where, from the ſoil, a crop of beans after wheat ſeems much more natural than a crop of turnips; and conſequently theſe ſhould be proportioned on ſuch a ſoil only to the bare neceſſity, and that neceſſity not increaſed by too large a ſtock of winter cattle; which, by extending the culture of turnips, or any winter vegetable over a larger ſpace of ground, than for which a ſufficiency of manure, ſo abſolutely eſſential to their produce, can be reaſonably procured, is ſure to diſappoint the very purpoſe, (and that frequently at a very great expence) which on a little ſcale muſt very advantageouſly have been procured. Where it can be afforded, cabbages, eſpecially in the drill culture, are certainly the propereſt for the intervals of beans; the ſame ſoil ſuitable to both. The cabbages planted at the very period when the beans begin to ceaſe vegetating, and their own vegetation promoted at firſt by the ſhelter that the beans afford, and afterwards, as their leaf decays, by gradually admitting on the young cabbage plants, in proportion to their ſtrength, the ſun and air.

It is with great ſatisfaction I ſee in Article 50th, on the repreſentation of Mr. Anstie, a premium offered

offered for the eſtabliſhment of Friendly Societies. The advantages of theſe are ſo ſtriking; the œconomy on which they are founded ſo reputable to the loweſt claſs, from whom they remove the ſtigma —that, ſure of a legal ſupport from the pariſh, they never will provide againſt their own misfortunes; and ſo beneficial to their ſuperiors, by whom this legal ſupport muſt be raiſed, and who conſequently ſhare in every ſhilling, that the eſtabliſhment of theſe clubs enables the members to be too independent to take; that I have been aſtoniſhed at not finding gentlemen in general, and every part of their families, (making it even a conditional agreement at the hiring of ſervants, that they ſhould be members of ſuch ſocieties) contributing by ſubſcription to their ſupport; but I have been more than aſtoniſhed at not finding any pariſh contributing out of the rates a trifle quarterly to the ſupport of the moſt obvious ſcheme, by which the maintenance of the poor, a burthen ſo juſtly every where complained of, can be reſtrained: even where many of thoſe ſocieties, ſo truly honourable, have been diſſolved by the ſudden and large calls of ſick members exhauſting that ſtock; and conſequently throwing them again on the ungrateful and improvident pariſh, that had neither generoſity enough to ſerve them, nor prudence enough to ſerve itſelf.

The

The very pariſh from which I write, in ſpite of my repreſentation, is of this folly a caſe in point; though in one inſtance of ſuch illneſs, it ſaved by ſuch a club nearly twenty pounds. What an univerſal ſubſcription, from every perſon at a certain age to thoſe clubs, conſidering how very large a portion would from ſituation draw no benefit from the collection, would do, may be at leaſt conjectured from a little pamphlet, publiſhed ſome time ago, by a Mr. PUGH, of *Wellingborough*, Northamptonſhire; who declares, that from the data afforded him in that pariſh, ſuch a general ſubſcription would not only anſwer the preſent poor-rates, but even probably afford an overplus for the aſſiſtance of poor families burthened with children, too young to earn any maintenance, and requiring all the care and time of the mother who ſhould contribute towards it. Inſtead therefore of ſtatute upon ſtatute, of perpetually deviſing new laws and new ſchemes,—a very melancholy ſign,—ſurely it would be but fair to try firſt whether the inefficacy complained of does not chiefly originate, and is not certainly increaſed, by the old laws we poſſeſs (preventive of every degree of vice, by reſtraining every place encouraging it) being never firmly and generally executed; nor to the good habits and cuſtoms that might ſtill be found openly and liberally ſupported

and

and promoted; and consequently, whether some security against these evils be not chiefly, if not entirely, within the hand that will not use it.

Yours, &c.

CHARLES ONLEY.

P. S. Erect Bridewells on the plan of that of *Wymondham*, Norfolk; unite the poor into hundred-houses of industry, like those in that county and Suffolk; strictly execute the laws against vagrancy, every sort of immorality, profaneness, licentiousness, and neglect of the Lord's-day; reduce the number of ale-houses to the bare necessities of every parish; support Sunday or similar schools on a cheap plan, towards forming the children of the poor by habit to some *little* knowledge, strict decorum, and *much* industry; contribute by general subscription to the support of the Friendly Societies, or poor men's clubs for mutual assistance, and the forming such in every town and village in the kingdom;—and *then* complain of a want of police, and of preventive justice—of security against a profligate commonalty, and an expensive poor, if you can!

It becometh none but children, when they possess all they want, out of weakness or wantonness, or both, to cry out for more. C. O.

 ARTICLE

ARTICLE VI.

On the Culture and Management of Rhubarb in Tartary; Method of using the recent Plant; curing the Root; Nature of its selenitic Salt, &c.

[By *A. F.* M. D. F. R. S.]

GENTLEMEN,

THE attention, which you have deservedly bestowed on the subject of RHUBARB, induces me to believe that a few additional observations, which have occurred to me since my last,* may not be unacceptable, particularly to those who wish to improve the culture and management of the plant in this country.

Mr. J. R. FOSTER, in his history of "VOYAGES "TO THE NORTH," very lately published, informs us, from the most authentick accounts, that at Suchur, a province subject to the Great Khan of Tartary, where the true plant flourishes in the greatest abundance, and from whence the merchants carry it all over the world, the country is rocky and mountainous, the soil red with a stratum of stone under it, sometimes boggy, being every where intersected with numerous rivulets.

* Inserted in the Society's Third Volume, Art. LVI.

At

At Kathay, and ſome of the more remote provinces, this root is held in no eſtimation, except for the diſeaſes of horſes, and for the purpoſe of common fuel. But at Suchur, where its value is better underſtood, its culture and management are duly attended to, and their method ſeems worthy of imitation in Great-Britain. The plant, in its native ſoil, flouriſhes luxuriantly, and the roots, when arrived at their full growth, are of an enormous ſize; the larger ones often meaſure three quarters of a yard in length, and are of the thickneſs of a man's body.

The roots are dug up in winter, before they put forth leaves, becauſe they then contain the entire juice and virtue of the plant; thoſe that are taken up in ſummer being of a light ſpungy texture, and unfit for uſe. The root being thoroughly cleaned, is cut tranverſely, and the pieces are placed on long tables, and turned carefully three or four times a day, that the yellow viſcid juice may incorporate with the ſubſtance of the root. If the juice be ſuffered to run out, the roots become light and unſerviceable; and if the roots are not cut within five or ſix days after they are dug up, they become ſoft, and decay very ſpeedily.

Four or five days after they are cut, holes are made through them, and they are hung up on ſtrings expoſed to the air and wind, but are ſheltered from the ſun-beams. Thus in about two months, the roots are completely dried, and arrive at their full perfection. The loſs of weight in drying is very remarkable, ſeven loads of green roots yielding only one ſmall horſe-load of perfectly dry Rhubarb! Concerning the age at which the roots are dug up, our author is ſilent; nor does this point ſeem yet to be clearly aſcertained. Some contend, that it arrives at its higheſt perfection in ſix or eight years; while others aſſure us, it ought to continue in the ground till the 10th, or even 12th year, before it acquires its full maturity.

Since the *Rheum Palmatum* has been cultivated in England, we have not heard of any uſe having been made of the recent plant. The Tartars, however, hold it in high eſtimation. Mr. THOUIN, ſuperintendant of the exotics at Verſailles, informs us, that the recent ſtem is converted into a marmalade, and is conſidered as a mild and pleaſant laxative, and highly ſalubrious. They prepare it by ſtripping off the bark, and boiling the pulp with an equal quantity of honey or ſugar. The leaves are employed in their ſoups, to which they impart an agreeable acidity, like that of ſorrel, which ranks in the ſame claſs with rhubarb. The ſeeds of the Engliſh plant contain the medicinal virtue of the root

in

in an eminent degree, as I have already hinted in my laſt,* and ſeem worthy of further inquiry. The ſelenitic ſalt, which I alſo there mentioned, has been but very lately diſcovered to be a conſtituent principle in rhubarb, and other aſtringent vegetables. The ingenious Mr. SHEELE pronounces it a combination of the acid of wood-ſorrel with a calcarious earth. Should this be found to obtain univerſally throughout that claſs of vegetables, it may help to enlarge our views concerning the nature of their aſtringent principle.

I am, Gentlemen, your very humble ſervant,

A. FOTHERGILL.

* See the Society's Third Volume, as above.

ARTICLE VII.

On a more ſpeedy Method of propagating Rhubarb.

By Mr. HAYES, Surgeon, at *Hampſtead.*

[Communicated by Dr. FOTHERGILL.]

GENTLEMEN, *Feb.* 15*th*, 1787.

NOtwithſtanding much has been written on the means of cultivating a very valuable root, the RHEUM PALMATUM; permit me to add my

 mite

mite to the general ſtock; as I think, by the following method, the plants may be raiſed in leſs time, and with greater certainty.

Having found myſelf diſappointed for many years back, in raiſing the rhubarb plants from ſeed, in the open borders of my garden, I was induced to try what ſucceſs I ſhould have by ſeparating ſome of the eyes or buds, which ſhoot out on the upper parts of the root, together with a ſmall part of the root itſelf, with ſome of the fibres to it; many of theſe may be ſeen, both in the ſpring and autumn, on plants of three or four years old. My ſucceſs was equal to my expectation; and all the rhubarb plants which I now grow, are raiſed after the above-mentioned method. I have juſt ſeparated twenty eyes or buds from a plant of four years' growth, which plant was itſelf raiſed the ſame way. The old plant is not at all injured, by taking the eyes from it, but is ſuffered to grow till it be ſeven or eight years old, or ſometimes longer, as the quality of the rhubarb, as well as the ſize of the root, will be much increaſed, if it lie in the ground till it be ten or twelve years old.

By the above method, I ſave a year in the growth of the plant; it is not in ſuch danger of being eaten by vermin as ſeed, nor ſo uncertain of its growing; it it is not ſo tender, neither does it need tranſplanting, or any other care than keeping the ground clear of weeds. I have not found any difference in the ſize of the roots thus raiſed, from thoſe which are raiſed from ſeed. I think my friend Sir WM. FORDYCE (whoſe views to enlarge ſcience in general are unceaſing) has remarked the uncertainty of the ſeeds of rhubarb coming up; and has pointed out, in the papers publiſhed by the Society of Arts, &c. in London, ſome means to render them more certain. Perhaps the above method may be thought preferable: if it ſhould not, pleaſe to accept the will for the deed; and believe me, that to be uſeful is the only deſign of, Gentlemen,

Your moſt obedient humble ſervant,

THOMAS HAYES.

ARTICLE VIII.

Account of a Plant of the Rheum-Palmatum, grown at Boreatton in the County of Salop.

[By a GENTLEMAN of that County.]

THIS plant, the ſixth year after it was ſowed, grew between the months of April (when the ſtalk firſt appeared out of the ground) and the middle

middle of July (when it was at its greateſt perfection) to the height of 11 feet 4 inches: when an obſervation was made on its growth, it grew in one day 3 inches, and in one night above 4: many of the leaves were above 5 feet long, the numerous branches all covered with bloſſom, and then with ſeed; in the latter ſtate by much the moſt beautiful. In October the ſeed was quite ripe, and the plant died down to the ground; the root was then taken up, and weighed 36lb. when clean waſhed and deprived of its ſmall and uſeleſs fibres. The method I took to cure it was as follows:—I pared off the outer rind, divided it with a ſharp knife into pieces of about an ounce weight, and then bored them through, ſtrung them on packthread, and hung them in the windows of an hot-houſe to dry. Some few I dried quick in an oven moderately heated, and did not find much difference; thoſe I attempted to dry in the ſhade became mouldy and uſeleſs; I ſhould gueſs the whole when dried, reckoning the ſuppoſed weight of thoſe pieces I ſpoiled by the laſt method, would have amounted to about 10 or 11 lbs. Of the refuſe pieces, ſuch as ſmall roots not thick enough to dry, I made a ſtrong infuſion in white wine, which I uſed with great ſucceſs in the dyſenteries of cattle; and ſome given occaſionally to poor people, when I thought it proceeded from cold.

The Method of Culture.

Sow the feed in your early cucumber bed, when a little of the firft heat is over. When the plants have got their third leaves, expofe them to the air; and when the feafon advances, remove the frame, leaving them in the bed:— in October or November take up the plants, the roots of which will be about half an ounce weight, and bury them under the mould prepared for the next year's hot-beds; the February following, plant them in an artichoke bed, which ought to be of the deepeft black garden mould, at leaft 2½ feet deep; after which they are to be treated in all refpects like artichokes, and about the fixth or feventh year taken up for ufe:—you may cultivate them from off-fets, when you take up a plant, but I think thofe raifed from feed better. I have never been able to obtain any feed from my plants fince that large one, the birds ever fince having conftantly deftroyed it before it was ripe.

Nov. 29th, 1783. J. S.

Article IX.

Obfervations on the Growth of fome Rhubarb, fent to the Society by George Poole, *Efq.*

Gentlemen, *Bicknoller, Somerfet.*

March 16, 1779. THE Rhubarb feeds were fown in the natural ground, and the 20th of March following the plants were removed from the feed-bed, and planted in a piece of garden ground

ground 44 feet long, and 22 feet wide, divided into four beds of about 5 feet each bed; holes 20 inches deep were made for each plant, and two ranks of plants in each bed, 3 feet afunder, (which I think is too near by a foot.) The ground was kept very clean from weeds, and every year in the month of October was dunged, the fame as for afparagus, and cleaned off again the fpring following.

Nov. 15, 1782, one of the borders of rhubarb was taken up; and of the roots, after they were cut, cleaned, and dried, there was left 54lbs. of good rhubarb; 50lbs. of which were fold in May laft to a druggift in London for £10.

In September 1783, the other three rhubarb borders were taken up, and produced 166lbs. of rhubarb of equal quality with the rhubarb herewith fent. The remainder of the rhubarb roots were produced from fome plants that grew in an orchard of very poor land. The rhubarb roots loft in drying about two-thirds. I am of opinion that the rhubarb going to feed (which will be in four or five years from fowing) leffens the weight of the roots, and tends much to its putrefaction; therefore it would be adviseable to cut off the feed-ftalks as foon as they appear. The crowns of the plant, when

when cut off and put into the ground again, will produce tolerable good rhubarb in four or five years, but not so large and plentiful as from seed plants.

The purging quality of this rhubarb is, I apprehend, not so strong as the foreign rhubarb; 30 grains of this rhubarb powdered being equal to about 20 grains of foreign rhubarb powdered. *Quere*;—If rhubarb seed mixed with clover seed, and sown with barley in deep rich lands, and permitted to remain four or five years, would not turn to much advantage, apprehending cattle of any sort would not hurt or eat the leaves? This is an experiment I purpose to make this spring. G. P.

Article X.

On the Danger of using of Lead, Copper, and Brass Vessels, in Dairies.

By Mr. Tho. Hayes, Surgeon, at *Hampstead*.

[Communicated by Dr. Fothergill.]

Gentlemen, *Feb.* 18*th*, 1787.

MANY eminent physicians have asserted, that butter is very unwholsome; while others equally eminent have considered it as not only innocent, but as a good assistant to digestion; and each

each have been ſaid to ground their opinions upon experience. Perhaps both may be right; and after all butter may be innocent or miſchievous, according as it contains many or few adventitious materials collected from veſſels, &c. uſed in the proceſs of making it.

I am led to theſe conjectures by obſerving, that in almoſt all the great dairies, the milk is ſuffered to ſtand in lead, braſs, or copper veſſels, to throw up the cream. The cloſeneſs of the texture of theſe metals, and their coldneſs and ſolidity, contribute to ſeparate a greater quantity of cream from the milk than would be done by wooden trundles, or earthen pans, both of which are alſo ſometimes made uſe of.

As I wiſh to eſtabliſh the poſſibility of the fact, that milk may corrode or diſſolve particles of the veſſels above-mentioned, and therefore be liable to communicate pernicious qualities to the butter, I beg leave to ſubmit my reaſons, from which I draw this concluſion; and if my opinion ſhould appear ſatisfactory to you, I make no doubt but you will do all in your power to diſcountenance the farther uſe of them; eſpecially as I ſhall point out others, which may be made, and will do as well for the dairyman's purpoſe.

Whoever

Whoever has been much in great dairies muſt have obſerved a peculiarly ſour, frowſy ſmell in them, although they be ever ſo well attended to in reſpect to cleanlineſs, &c. In ſome, where the managers are not very cleanly, it is extremely diſagreeable, owing moſtly to the corrupted milk.* In ſome, too, from the utenſils being ſcalded in the dairy; and in others, from a bad conſtruction of the building itſelf, the want of a ſufficient circulation of air, water, &c.; but in all, a great deal of the lighter or more volatile parts of the milk fly off from the ſurface of the pans, and furniſh a great quantity of acid effluvia to the ſurrounding air and ceiling, and which is again depoſited on every thing beneath it, and of courſe often on the veſſels after they have been put by clean, in the intervals of their being out of uſe. This may be obſerved to give a dull ſort of appearance to braſs and copper, as if you had breathed upon them; for if you rub your finger lightly over the veſſels, you will have both the taſte and ſmell of the metal.

It alſo happens ſometimes, that after the veſſels earw aſhed, they are not carefully rinſed, nor perfectly dried by the fire; ſo that ſome of the milk, &c. is left on the ſurface of them, which may

* See a very ingenious paper by Mr. HAZARD, in the Third Volume of the Bath Society's Papers, on making butter.

diſſolve

diffolve the metals, either by its animal, oily, or acefcent qualities.

This is not the only way, nor the worft, by which the butter may become impregnated with mifchief. The greater the quantity of cream that is thrown up from the milk, the larger profits accrue to the dairyman; therefore he keeps it in the veffels as long as he can, and it is frequently kept 'till it is very four, and capable of acting upon them; if they are of lead, a calx or fugar of lead is produced; if of brafs or copper, verdigrife.

It is true, the quantity cannot be very great; but this will depend upon the degree of fournefs, and length of time which the milk ftands:—but independent of the acid, the animal oil in the cream will diffolve brafs and copper.

That an acid floats in the atmofphere of a dairy, may be proved by placing a bafon of fyrup of violets for a little time, and it will be found to turn red.

Then, gentlemen, if I am right in my conjectures, as I am perfuaded I am, from the innumerable experiments and obfervations which I have made to fatisfy myfelf of the fact, and which would be trifling

trifling with your time and patience to relate here, —may not the reputation of the wholſomeneſs or unwholſomeneſs of butter, depend upon, or be owing to ſome of the above cauſes? And may not many a caſual, nay, obſtinate complaint originate from the ſame ſource, which the phyſician may have in vain laboured to account for? Butter is found very frequently to occaſion much diſorder to very weakly, delicate, and irritable ſtomachs; yet theſe ſtomachs will bear olive oil:——this cannot be therefore accounted for from the oleoſe parts, but may from the metallic impregnation.

I will not contend, that all the ill effects attributed to butter are cauſed by the mineral particles which it gains by the means above ſtated. I only inſiſt that it is poſſible, and indeed very probable, and that it may in conſequence do frequent miſchief; and that when butter is free from theſe particles, it is not ſo unwholſome as ſome have aſſerted; though when it does contain them, it is found to diſorder very tender perſons.

To enlarge upon the ſubject, or attempt to explain the many ways and how a very ſmall portion of the above metals may prove injurious to the human frame, in ſome particular conſtitutions, will be only to repeat what has already been ſaid

by

by abler writers.* Some will perhaps ſay that my ideas are very far fetched, and others that my opinions are ill-founded; but I truſt whoever has read the induſtrious reſearches of the very learned Sir GEORGE BAKER, on the effects of lead, and the melancholy caſe of a young lady having died from eating pickle ſamphire very lightly impregnated with copper, and which others ate without being diſeaſed, as related by the ingenious Dr. PERCIVAL, will receive my opinions with leſs objection. If I have erred, I have done it in honourable company.

If you think with me, gentlemen, I hope you will have ſome influence over the dairymen, to induce them to change their utenſils, as very commodious veſſels may be made of caſt-iron equally well fitted for their purpoſes, which will not prove expenſive, and will be more innocent and cleanly. But if they continue in the habit of uſing thoſe above-mentioned, after they are informed of the bad tendency of them, they muſt be guilty of a great breach of moral duty, and highly blameable, both in a religious and political point of view.

* See Sir GEORGE BAKER's papers on the effects of lead, in the Medical Tranſactions;—Dr. PERCIVAL's paper in the ſame;—and Dr. FALCONER alſo on copper veſſels.

I beg

I beg pardon for thus treſpaſſing upon your time and patience, but I truſt you will excuſe my errors, and alſo my prolixity, ſince it has for its object the preſervation of the health and happineſs of my fellow creatures.

I am, Gentlemen,

Your moſt obedient humble ſervant,

HAMPSTEAD, THO[s] HAYES.
Dec. 5th, 1786.

ARTICLE XI.

On the Culture of Rape or Cole Seed.

THE proper time to ſow Rape, broad-caſt, is the month of June; the land ſhould, previous to the ſowing, be twice ploughed and well pulverized; when about two pounds of clean ſeed will ſuffice for every acre, which ſhould be equally caſt upon the ground with the two fore-fingers and the thumb; for if it be caſt with all the fingers, it will come up in patches, and be the means of waſting ſeed. When the plants appear, if they come up too thick, a pair of light harrows ſhould be drawn, length-ways and croſs-ways, over the land; this will equally thin them, and when the plants (that the harrows have pulled up) are withered, the ground ſhould be rolled, and a few days after the plants may be ſet out with a hoe; 16 or 18 inches is the diſtance proper for each plant to flouriſh in.

In the North of England, the farmers pare and burn paſture lands, and ſow them with rape after one ploughing; which crop commonly ſtands for ſeed, and will fetch from 25l. to 30l. per laſt,* for the purpoſe of making oil. Poor clay, or ſtone-braſh-land, will frequently produce from 12 to 16 or 18 buſhels per acre, and almoſt any freſh or virgin earth will yield one plentiful crop. Many in the Northern Counties, by cultivating rape, have been raiſed from poverty to the greateſt affluence. The ſeed of it is ripe in July or the beginning of Auguſt; and it is ſurpriſing to ſee with what avidity people flock to a rape threſhing (as it is called in the North). It is an abſolute feaſt; a violin is conſtantly played in the field, while the buſineſs is performing; the beſt of proviſions are procured, and a rural dance concludes the evening's diverſion: mirth and good-humour mark the happy countenances of all who aſſemble; and thoſe who are or are not invited, equally partake of the proviſions and pleaſures of the day.

The rape is cut by men with hooks or ſickles, and ſpread thinly on the ground to dry, and when

* A laſt is two loads=ten quarters=eighty buſhels.

it is found in order for threshing, the neighbours are invited, who endeavour to render themselves useful. A number of cloths are in readiness, for the purpose of carrying the seed to the threshers; who perform their business on a large cloth in the middle of the field—and here the fidler displays his skill.

The seed is put into sacks and conveyed home; and a field of 20 acres or more is completely harvested and threshed in one day upon the spot; nor will rape admit of being carried from the ground in the pod, as it must be perfectly ripe, and would therefore shed or scatter; the straw the farmers burn, and dispose of the ashes, which are allowed to be as valuable as the best pot-ashes.

Rape that is suffered to stand for seed, will very much impoverish old arable land; but pasture lands, that are previously pared and burned, will bear two or three good crops of corn after, without manure.

The price for paring and burning, varies in different counties; in Yorkshire, and still further north, it is performed for from 10s. to 15s. per statute acre, and in the south and west parts of England, the price is from 16s. to one guinea.

Rape is an excellent food for ſheep, and for this purpoſe it will anſwer well on arable land; but it ſhould be hoed and ſet out as before directed, and it will be the ſtronger, and produce a much heavier crop, if it be looked over a ſecond time, and the earth be drawn round the ſtems; and ſhould there appear any places where the crop has failed, it will be right to draw plants where they are found too thick, and plant them in the bare places; by which means a more general and equal crop may be expected; and that which is tranſplanted, will be ſuperior to any of that which has never been removed.

The writer has experienced the good effects of tranſplanting rape, and begs leave ſtrongly to recommend it; he adviſes a plot of ground of about a rood, to be ſown in the middle of June; this will produce plants enough for ten acres, which may be planted upon land that has previouſly borne a crop of wheat, provided the wheat is harveſted by the middle of Auguſt: one ploughing will do for theſe plants; the beſt of which ſhould firſt be ſelected from the ſeed plot, and be planted upon ridges at leaſt two feet aſunder, and ſixteen inches apart in the rows; they may afterwards be horſe or hand-hoed, and the earth ſhould be drawn round their ſtems; and in the ſpring of the year this crop may be be fed off with sheep, when very little other green fodder is to be found, or the leaves might be gathered and given to oxen or young beasts; from the same stems fresh leaves would sprout again, and these might be fed off by ewes and lambs, time enough to plough the land for a crop of barley or oats; but it must not be forgot that planting rape upon land the beginning or middle of July, would be the most advantageous as to the crop of rape, as the leaves might be then fed off in the autumn season, and they would still produce other leaves anew in the spring; and this method of early planting might be adopted where pease or beans had been gathered green, and sent to a market, or where any kind of pulse or green fodder had been fed off the preceding spring.

The expence of planting rape varies according to the price of labour in the different counties; but the most general price, where rape is planted on ridges two feet asunder and sixteen inches apart in the rows, is 2s. 6d. or 3s. per acre; but where every plant is to be equidistant, or 16 inches every way apart, on a flat surface, 3s. 6d. or 4s. per acre is about a fair price, provided in either case the grower finds a woman or boy to draw the plants from the seed plot, and to drop them before the planter. When they are thus planted, they may be hand-

 hoed

hoed and earthed for 4s. 6d. per acre; but where rape is ſown broadcaſt, few will be found to hoe and ſet the plants out at equal diſtances, and earth them for 7s. per acre, nor will the plants ever flouriſh equal to thoſe which are planted; therefore it is obvious which method is to be preferred.

The practice of ſowing rape and turnips (if they are afterwards ſuffered to ſtand together) is by no means commendable, as it is not poſſible to hoe or ſet out both ſorts to advantage. If the rape flouriſhes beſt, the turnips ſhould be deſtroyed, and *vice verſa*; otherwiſe the crops would injure each other, as the lateral fibres of the rape would be prevented from expanding, if ſurrounded with turnips; and in froſty weather the water or dews would drop from the leaves of the rape on to the turnips, and totally rot and deſtroy them. It is a common practice with farmers to feed ſuch crops with ſheep, and afterwards to plough up the land for wheat; on which account it appears unneceſſary to them to hoe or ſet out either crop; but let them conſider, if it be right to ſow turnips and rape, it is equally to their advantage to hoe them; as the hoed crop will produce a burden at leaſt five times larger than the crop that is not hoed, this will amply defray the expence, and it ſhould be a conſtant rule always to deſtroy the weakeſt crop.

Thoſe

Thoſe who look for an immediate profit, will undoubtedly cultivate rape for ſeed, but it may anſwer perhaps better in the end to feed it with ſheep; the fat ones might cull it over firſt, and afterwards the lean or ſtore ſheep might follow them, and be folded thereon; if this be done in the autumn ſeaſon, the land will be in good heart to carry a crop of wheat; or where the rape is fed off in the ſpring, a crop of barley might follow; in either caſe rape is valuable to the cultivator; and when it is planted and well earthed round the ſtems, it will endure the ſevereſt winter; but the ſame cannot be advanced in favour of that which is ſown broadcaſt.

I flatter myſelf the foregoing obſervations will prove acceptable to the ſociety, and uſeful to gentlemen farmers in general, who may wiſh to cultivate rape either for ſeed or fodder.

I am, Gentlemen,

Your very humble ſervant,

Stoney-Littleton. J. HAZARD.

ARTICLE

Article XII.

[*The following Letter was drawn up for the more immediate Use of the* "Odiham Agriculture Society," *who had done the Writer the Honour of electing him an Honorary Member*;—but being thought generally beneficial, *its communication is extended.*]

ON THE MEANS OF PRESERVING APPLE BLOSSOM AND ORCHARDS FROM INJURY.

Sir, *Beerferris, near Taviſtock, Devon.*

I Do not recollect that yours is a noted cyder county, yet I take it for granted, ſome gentlemen have orchards, and the rarer the more valuable. This pariſh, which is my ſummer reſidence, abounds with orchards and cherry-gardens; the orchards, by their bloſſoming this ſpring, promiſed a much larger quantity of apples than they will actually produce; not occaſioned, however, as the farmers here imagine, by the froſty nights of the firſt and ſecond of laſt month, but by the ravages of an uncommon number of *inſects*, which have been produced this ſeaſon from a ſpecies of black flies in particular, which depoſited their eggs in the apple-bud, or bloſſom, at its firſt opening; from which eggs were generated the maggot inſects, which

which by feeding on the heart of the bud or bloſſom, ſoon occaſioned it to drop, contract, and cloſe itſelf into the form of a cup, of a brown red colour, reſembling that of a dry dock-leaf, (unleſs this was originally cauſed by the bite of the fly, when ſhe depoſited her egg there) ſo as to afford a ſafe nidus for the young inſect, and ſufficient nouriſhment to ſupport it, until full grown in that ſtate, and needing no longer protection there; when it decamps, and the bloſſom being deſtroyed, at laſt falls off—whereby a plentiful bloſſoming is likely this year to produce (as is often the caſe) a ſcanty bearing.

I have within a few weeks paſt opened ſome ſcores of thoſe ſhrivelled bloſſoms, and ſcarce ever failed of finding a maggot inſect (ſome much larger than others) ſafely incloſed within its natural neſt; though in ſome inſtances I found it had decamped, after having exhauſted its nutriment, and the decayed bloſſom was ready to fall off with the ſlighteſt touch.

Having thus given you a plain conciſe account of the evil, permit me, Sir, to point out what I conceive will be an *effectual remedy*, or rather preventive; and will likewiſe contribute to the fertility of the ſoil, the proſperity of the trees, the future produce of apples, and the goodneſs of the paſture.

When

When the winged infect tribe firft begin to appear, (which fome conceive, and not improbably, to be by an eaft wind bringing fome forts at leaft of them over from the continent) I would recommend fome heaps to be made of the fward or fpind, in the nature of denfhiring or burnbaiting, or heaps of long dung, wet ftraw, weeds, or any other like matters, at different intervals all around, *i. e.* on every fide, and likewife fome in different parts of the orchard. If an eaft wind blow, fet fire to fome of the heaps on the eaft fide, and fome within the body of the orchard; if a fouth wind, then on the fouth fide; and fo occafionally on different fides, as the wind may happen to vary; but always on that fide from whence the wind happens to blow, fo that the fmoke from the fmothering of the heaps may blow through and fumigate the orchard for fome weeks. The expence attending which will comparatively be very trifling, but its confequences and beneficial effects very great, as it will effectually prevent the infect fiy not only from depofiting its eggs, but even from approaching, or at leaft continuing long in fuch a noxious fituation, whereby the bloffoms and fruit will be preferved from fuch ravages, and the heat and afhes of the fmothering heaps will likewife contribute to the fertility of the foil, the fweetnefs of the pafture, and the growth and vigour of the trees

for

for future bearings; thus destroying *moss* probably better than by any other means, and counteracting the effects in some measure of cold and blighting winds, and such late frosty nights as those of the first and second of last month.

I am, Sir, your's, &c.

June 3, 1786. C. GULLETT.

P. S. It seems superfluous to add, that the same process is applicable, and promises to be equally advantageous, to *all other fruit trees*, if suitably adapted in point of time and other circumstances.

Article XIII.

Success of an Experiment of fumigating an Orchard.

Sir, *Nov.* 23, 1786.

THE foregoing letter was drawn up on the day it bears date, the 3d of June last, and was intended to have been communcated at that time, but a diffidence of publishing my theoretical ideas, unaccompanied by facts, prevented; and accordingly it hath lain by ever since, and so would have continued but for the following information.

Since

Since this idea firſt occurred to me ſome years ago, I have occaſionally mentioned and recommended it in ſtrong terms, to very many gentlemen farmers, cyder growers, and others, both in the eaſt and weſt parts of Devonſhire, and in Cornwall; but ſo little are recommendations of any innovation or improvement heeded, eſpecially by the generality of farmers, that I have never heard, till very lately, of any one having put it in practice, and that one inſtance occaſions my troubling you with it, as the ſucceſs of this farmer may induce others to adopt it, until *the fumigation of orchards*, in bloſſoming ſeaſon eſpecially, becomes general; and a moſt precarious crop is thereby rendered as ſure and certain as a crop of corn.

I have been well informed then (at firſt by a gentleman farmer, with great glee and ſatisfaction, to whom I had ſome years ago recommended it) of a farmer in the eaſtern part of Devonſhire, who this ſpring determined to give a fair trial to the ſucceſs of fumigating orchards; and in order thereto, made choice of one orchard to be fumigated, leaving another (ſimilarly ſituated and circumſtanced) unfumigated. The event of which was, that the *fumigated* orchard had a very large and plentiful bearing of apples, to his no ſmall emolument; which he attributes, and I believe, very juſtly, to

this

this ſumigation alone; while the *unſumigated* orchard and neighbourhood had ſcarce any apples at all.

Your moſt obedient ſervant,

CHRIST. GULLETT.

ARTICLE XIV.

Means of inſuring full Crop of Turnips.

SIR, *Exeter, Nov.* 24, 1786.

THE very great failure of Turnip crops, ſo generally and almoſt annually complained of, in different parts of the kingdom, have occaſioned a variety of recipes to be publiſhed, but which on trial too frequently fail. The turnip ſeed itſelf ſeldom fails to vegetate, (eſpecially if ſown juſt before rain falls, and as ſome ſay, if left to itſelf, neither harrowed nor rolled, when ſown dry and rain ſoon follows) but the damage is occaſioned in its ſubſequent ſtate of leafing, by the voracity of *inſects.* To prevent this, ſeems the grand *deſideratum* in turnip agriculture; and a moſt deſireable one indeed it certainly is, as the loſs in turnip crops this ſummer, in *Devonſhire* alone, is calculated at upwards of 100,000l. The uſe of ſteeps, &c. for the ſeed, however, appears to me very little likely

to

to produce this effect. In lieu of all which, I beg leave to recommend the adoption of the following idea, which hath occurred to me, in the courfe of writing my preceding letter dated yefterday; of the certain and never-failing effects of which, in the moft fatisfactory manner, I will not hefitate to exprefs an abfolute confidence.

The damage in orchards is done by infects, fo it is here, though of a different fpecies; the former effected by the infect in its maggot ftate, the latter by the fly: but as the fly is the original caufe in both, fo the fame means are applicable in both cafes; and therefore let the farmer make himfelf mafter of the method I have recommended for the fecurity of apple bloffom, and very little more need be added here.

If the turnip ground be fpaded and burnt, or the weeds, &c. burnt without fpading, the fumigation thereby may fuffice to chafe fuch of the infect winged tribe from thence as are then there; but in all cafes, when the field is ploughed and ready for fowing, let heaps be made at different places and intervals round by the hedges and boundaries of the turnip ground, and fome few fcattered through the field, in the fame manner as directed for the orchards. Then, as foon as the feed is fown, let the

heaps

heaps on the *windward ſide*, and the ſcattered ones, be lighted and kept ſmothering during the continuance of the wind in that quarter; the leſs the fire and the more the ſmoke, the better. Should the wind happen to ſhift, thoſe heaps on the quarter it ſhifts to muſt then be lighted, and kept ſmothering in like manner; ſo that during the growth of the tender turnip leaf, and until it becomes rough and out of danger, this fumigation and ſmoke over and acroſs the field muſt be continued from one quarter to the other; which, I venture to aſſert, will effectually deter and prevent any winged inſect tribe from approaching the turnip ground; nay more, if there already, it would moſt compleatly drive them from thence, as ſuch delicately formed inſects (which can only feed upon the moſt tender leaf) would be ill able to continue long in ſuch a ſmother of fire and ſmoke. The conſequence is obvious and certain, that if the fly be kept from approaching the field, *the turnip crop is ſafe*—and few, I believe, will diſagree with me, that *prevention is better than remedy*.

I am induced to be the more ſanguine of the ſucceſs of this method, from the great ſucceſs, which on many trials attended a ſimilar idea of mine, for the preſervation of cabbage plants from caterpillars by means of elder buſhes, which was inſerted inferted in the Sixty-fecond Volume of PHILOSOPHICAL TRANSACTIONS," about the year 1773; from whence it was copied into the *Annual Regifter*, magazines, and periodical and other publications.

In order, however, to be fure not to fail of obtaining the full effect and utmoft fafety, (though it feems a work of fupererogation) let me, in addition to the above fumigation, recommend it to the farmer, who harrows or brufhes in his turnip feed, to add to his harrow or thorn-bufh, a bufh of ftinking elder, [SAMBACUS] the dragging which over the field will leave fuch a difagreeable fmell and effluvia behind it, as might, and would, I think, be fufficient alone, without fumigation, (as was the cafe with butterflies and of the caterpillars above alluded to) but when united with fumigation, no farmer who adopts this recommendation, I will venture to promife him, need be at all uneafy in future about the fuccefs of his turnip or *any other* vegetable crop; and fuch as know it, and refufe to adopt fo cheap a remedy, deferve little pity, if their crop is devoured by the infects.

I could proceed to apply a fimilar remedy for *wheat crops* from being damaged by the yellows and other infects; of which I have difcovered upwards of forty living ones, inclofed within the hufk of of a ſingle *grain* of wheat, as ſtated in the above caterpillar hiſtory, to which I refer; but *that* muſt be the ſubject of a future letter.

If it ſhould be aſked,—What would become of thoſe moſt numerous tribes of prolific inſects, if this method ſhould become general? I moſt ſeriouſly anſwer, that I firmly believe not a thouſandth part of them would trouble us in a few years hence. And in a *philoſophical* light, I cannot but conſider that ſuch very general fires and fumigations throughout the kingdom, would tend very materially to the rarefaction, purifying, and improvement of the ſtate of the atmoſphere, and procuring healthy ſeaſons.

As I conſider this an object of conſequence, I loſe no time in conveying theſe my ſentiments to you for the good of the publick, which is the object of your laudable inſtitution, as well as of my amuſement at leiſure hours.

Your moſt obedient ſervant,

CHRIST. GULLETT.

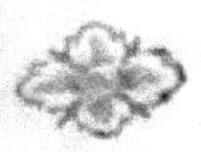

Article XV.

On the Culture of Mustard;—Remarks on the Trifolium Alpestre;—on the Necessity of Change of Artificial Grass Crops, &c. &c.

SIR, *Stifted-Hall, Essex, Dec.* 14, 1786.

IN order to anſwer, as ſatisfactorily as I could, your enquiry about the cultivation of muſtard, I ſent for one of the *ſeed-cultivators*, of which there are ſeveral in this neighbourhood; eſpecially about *Coggeſhall* and *Kelvedon*. Theſe men hire, at very advanced rents, a few choice acres of the farmers; cultivate them, merely for the ſeeds of various ſpecies of pulſe, roots, herbs, and even ſome flowers, with the utmoſt gardening neatneſs; and ſpeculate upon the chance of produce, ſale, and price, in which they have ſome little portion of the acuteneſs of an alley-broker,—like him fluctuate from very great profits to ſome loſſes: but if they can ſtand out contingencies, ſucceed upon the whole amply enough.

The white muſtard requires rather an heavy ſoil, which muſt by tillage be brought into a nice mould; muſt be ſown in March at one buſhel an acre; be always twice, and frequently three times, hoed, and ſet

ſet out at about ten inches plant from plant. The crop is reaped in Auguſt, and leaves the land in ſufficient tilth for any crop of other grain or corn that may be choſen to follow it: the *medium* produce three quarters per acre, and the medium price 10s. per buſhel. Muſtard never follows muſtard; but may be ſown on the ſame land again in the third year. The firſt hoeing is worth 4s. the ſecond and third 3s. per acre.

I never ſaw the marle-graſs you mention; but obſerve, you affix the ſame Latin name to it as you did in my letter [Article XXXIV.] inſerted in your laſt volume—trifolium *alpeſtre*, to what I there called cow-graſs; and which is, trifolium *purpureum pratenſe*; and from the trial I there mentioned to you, appears to me a moſt valuable ſpecies. The trifolium *alpeſtre* is, I apprehend, the *real* cow-graſs; though the other is, at the ſeed-ſhops, ſold under that name. Your ſociety therefore, if they have acquired any quantity of the ſeed of the *real alpeſtre*, which has been thought to be *particularly rare*, will bring a very great acquiſition to agriculture in one of its moſt eſſential points.

Here common clover frequently, through the accident of ſeaſons, rather than quickneſs in cropping with it, will fail. In Norfolk, where it has ufually come over again at every fourth crop, this failure, from the land being furfeited with it, has been in many inftances fo great, that a very confiderable farmer there lately informed me, that he and many of his neighbours had lately under their barley fown rye-grafs and trefoil; and in only the *next* rotation of barley fown the common clover; and fo on alternately: for not lefs in artificial grafs than corn, change and variety, under the *common hufbandry*, is effential to the produce.

The very barley in Norfolk, probably from the fame caufe, has of late years, it is faid, degenerated in finenefs, befides varying more in the rotation of crops.

Our farmers defend the mode of the *whole year's fallow*, on the principle of its guarding againft fuch furfeit; and thus by the delicacy of the fowing tilth it neceffarily produces, making the proper bed for minute grafs-feeds, and giving them the beft fecurity againft the failure of their clover. They frequently mow it for hay, and then feed it, or for the chance of the latter, (a very dubious, though fometimes a moft profitable one) feed it off firft early; but very feldom let it remain a fecond year. I *conjecture*, that where it muft neceffarily come in quick rotation, it fhould never ftand but one year,

and

and in that be *conſtantly fed*; it may then alone be regarded as a *meliorating* crop, but otherwiſe as a *wearing* one; and a diſtinction of it is thus made in ſome of our leaſes. In the one inſtance the land is *probably* ſurfeited, by puſhing forth into full bloom ſuch a ſtrong crop, and continued too perhaps to a ſecond year; in the other, it is cheriſhed by its paſturage, and improved by its being early turned in. When intended for merely one year's feed, any of the other artificial graſſes may, to ſecure a plant, be ſown with it.

My preſent wheat is on the rye-graſs ley, mentioned in your third volume, [Article XXXIX.] and appears as perfect as on one of clover. My field for early feed of next ſpring is alſo rye-graſs *thickly ſown*, and, as a little trial, I have ſown with it, in one portion, ſome trefoil; in another, the perennial white; and in the third, perennial red clover, commonly termed *cow-graſs*.

However congenial to the ſoil of a farm any peculiar ſpecies of corn, pulſe, or vegetable, may appear; the chance of a ſecure crop from them, in a large ſcale, under, as I ſaid before, the *common huſbandry*, will in general be in a proportion to their not being ſown too often:—the variableneſs of our climate corrects ſo frequently, or brings to ſuch a fort of equality, the variety of one foil, that in proportion to the neceffities of his ufual ftock, the conveniency of culture, and the benefits of rotation, as corn, pulfe, or artificial winter herbage, and fpring grafs, may, though not in equal, yet in fome proportion, be advantageoufly raifed on foils, in appearance improper for one or the other of them, by every attentive farmer; and his profit in a long leafe, on this broad bafis of rough agriculture, be at the end more fecure, though occafionally it cannot be fo great. I am, refpectfully your's,

CHARLES ONLEY.

Article XVI.

Account of the Cultivation and Produce of a Crop of Buck-Wheat.

[By a Gentleman Farmer, to the Secretary.]

SIR,

AGREEABLE to your requeft, I fend you the following account of the cultivation of 11¾ acres of Buck-wheat.

The inclofure No. I, containing feven acres, was a deep, friable, fandy loam; its afpect a gentle flope towards the fouth. In 1784, it produced a very poor

poor crop of wheat, being ſmutty and very full of weeds. The wheat-ſtubble was ploughed up in November following, and a good coat of dung ſpread over it. In this ſtate it remained till the middle of laſt May, by which time it became full of couch and other noxious weeds.

The ſloven of a tenant then giving up his leaſe, I had the furrows turned back, then cut acroſs, well dragged and cleaned with a couch-harrow—an excellent implement for diſpatch—a horſe, man and boy, doing in a day as much work as a dozen people with rakes. Next I gave a deep ploughing lengthways, harrowed and couched again; which brought the land in fine tilth, and exceeding clean. I finiſhed ſowing about the 1ſt of June, and harrowed and rolled afterwards. The buck-wheat came up about two inches high, regular and even, when the long drought commencing checked its growth, and caſt a ſickly yellow hue over the whole, particularly under the hedges, where it lay as flat as if cut off, a conſiderable quantity being burnt almoſt to a coal. In this condition it continued languiſhing for about ſix weeks, when a fine ſhower produced an amazing alteration; it immediately reared its drooping head, and tillered out into branches ſix or eight in general on a ſtalk.

The

The appearance was now agreeably changed, the whole field being covered with a moſt beautiful white carpet; and an innumerable multitude of bees buzzing in every part, preſented a ſcene truly romantick. From this time through the ſummer, it continued matting together, growing and bloſſoming till full a yard in height, promiſing a fine crop. To appearance the ſeed kerned remarkably well, ſeveral people who went to view it, concluding there would at leaſt be a load upon an acre. The ſecond week in September, began mowing, and turned about two acres; but perceiving, as the haulm was ſo long and ſtalky, that in turning a vaſt quantity would be ſhed out, I ordered a couple of men to go to two ſwarths, aud take gently up a ſmall quantity, and place each parcel againſt its fellow, between the ſwarths, the ſeed uppermoſt:—by this means the ſun and air circulated freely through the haulm, drying it faſter than turning would have done. Once ſetting up ſufficed, beſides the advantage of preſerving the ſeed. Before it was dry enough to harveſt, the wet weather commenced, which obliged us to keep it frequently moved to get it in order, as well as to prevent the ſeed growing; of conſequence this occaſioned the loſs of a conſiderable quantity, and delayed houſing it till the beginning of October.

The

The buckwheat left this piece in exceeding fine condition for wheat, not a weed or ſcarce even a blade of graſs was to be ſeen throughout the whole. Obſerve:—An acre of this piece was ſown with turnip-rooted cabbage, for an experiment; which failed, though the plants came up thick enough; yet after hoeing and weeding, they made ſuch little progreſs, that the weeds got the aſcendency and choaked them. Deduct likewiſe about forty lug of grubbed hedge-rows, and a road running up the middle, which is included in the ſeven acres.

The fields No. II. and III. contained, the one four, the other two acres, ſomewhat ſtronger land than No. I.; the expoſure a gentle deſcent towards the eaſt, except about an acre lying hollow, which is ſubject to be overflown after heavy rains. Theſe pieces were oats in 1784. As ſoon as No. I. was ſown, I gave them one earth; the weather being likely to continue dry, I harrowed and ſowed directly. A little of the ſeed came up between the ridges, where the ground was ſomewhat moiſt; the reſt remained as dry as when ſowed till July, when the rain that occaſioned No. I. to flouriſh ſo much, cauſed this to make its appearance; but it was very irregular, ſtraggling and weak. It ſoon came into bloom, tillered but little, and growed ſlowly, ſcarce attaining above a foot or 18 inches in height, and fore-

boding

boding but a lamentable crop. However, it was ſome amends to have a fine ſeaſon for harveſting, which was finiſhed by the 8th of October. The land was in tolerable order for the enſuing wheat crop, though nothing to compare with No. I. it not ploughing ſo free and mellow; beſides the couch was got up, particularly in No. III. which is the wetteſt of the two pieces.

The expences of cultivation have been very heavy, as you may judge from the neglect of the late occupier, and its being at a diſtance from home, on which account I have been obliged to hire for every thing. However, I have endeavoured to ſtate accurately every expence which ought to be charged, as follows:—

Inclosure No. I.——*Debtor.*	£.	s.	d.
To ploughing 5¾ acres in Nov. 1784, at 7s.	2	0	3
Four horſes and 2 men, dunging 2 days, at 9s.	0	18	0
To ploughing 3 times in May 1785, at 6s. -	5	3	6
Dragging, 4 horſes and a man, 1 day - -	0	8	0
Couching and carting off with 2 horſes and 2 men, 2 days - - - - - -	0	10	0
Nine buſhels of buckwheat, at 5s. - -	2	5	0
Sowing and rolling - - - - - -	0	4	0
Rent - - - - - - - -	5	15	0
Tithe compounded for - - - -	1	1	0
Poor-rates - - - - - -	0	10	0
Two men 3 days mowing, at 2s. - -	0	12	0
	£.19	6	9

	£.	s.	d.
Brought up - -	19	6	9
Two men at intervals turning, carting, &c. a fortnight each, at 1s. 4d. - - - -	1	12	0
Two children 5 days each, turning at 3d. -	0	2	6
Threſhing 21 ſacks, and 1 buſh. buckwheat -	1	0	3
	£.22	1	6

PER CONTRA.——*Creditor.*

By 21 ſacks and 1 buſhel buckwheat, at 16s.	17	0	0
Value of haulm for litter, &c. - - - -	3	0	0
Loſs to balance - -	2	1	6
	£.22	1	6

INCLOSURES No. II. and III.

In June ploughing 6 acres, at 7s. - -	2	2	0
Dragging and couching with 2 horſes and 2 men one day - - - - - - -	0	9	0
Two days 1 man raking, at 1s. 2d. - -	0	2	4
Nine buſhels buckwheat, at 5s. - -	2	5	0
Sowing and rolling - - - - - -	0	4	0
Rent - - - - - - - -	6	0	0
Tithe - - - - - - - -	1	1	0
Poor-rates - - - - - - -	0	10	0
Mowing 6 acres, at 1s. 2d. - - - -	0	7	0
Three children turning 6 days each, at 3d.	0	4	6
Carting, with 2 horſes and 2 men, 1 day -	0	6	0
Threſhing 12 ſacks, at 1s. - - - -	0	12	0
	£.14	2	10

PER CONTRA.——*Creditor.*

By 12 ſacks buckwheat, at 16s. - -	9	12	0
Value of haulm - - - - - - -	2	0	0
Loſs to balance - -	2	10	10
	£.14	2	10

According to the above computation, I think the cultivation of buckwheat to be of great conſequence to the community as a ſubſtitute for a fallow, as no fallow can exceed, or hardly equal, the piece No. I. for condition; though as a crop it has fallen greatly ſhort of my expectation, yet as it ſtands, I think having a fallow of twelve acres of ground for about 5l. expence, very reaſonable.

I finiſhed ſowing with wheat on one earth the 6th of November, the expence of putting in as under-mentioned:

	£.	s.	d.
To ploughing 12 acres, at 6s. - -	3	12	0
Five days at harrow, at 5s. - -	1	5	0
Nine bags ſeed-wheat, at 11l. 15s. per load -	10	12	6
Sowing - - - - -	0	4	0
Two days ſtriking furrows - -	0	10	0
	£.16	3	6

N.B. I have ſold about two quarters of buckwheat for fatting pigs, at 4s. per buſhel, for which reaſon I made that the average charge. An acquaintance of mine has ſown one field alternately wheat and buckwheat for three or four years paſt; rolling down the buckwheat, and then ploughing it in. He informed me that his laſt crop of wheat yielded four quarters per acre after the buckwheat.

N.B. The

N. B. The above account of cultivating buckwheat appearing very extraordinary and unnecefsarily expenfive, we fhall fubjoin the remarks made on it by a very ingenious gentleman farmer from Suffolk, who has long cultivated this grain in a very different and much more profitable manner.

" *To the* COMMITTEE.

" GENTLEMEN,

" THE letter on the cultivation of Buckwheat, on which you defire my fentiments, appears to have been written by a gentleman totally unacquainted with the management of that particular grain, and not fufficiently verfed in the true principles of agriculture.

" His foil was of the beft quality, and advantageoufly fituated;——but what rational, practical farmer would, after one earth in November, immediately fpread a large coat of dung on a foul wheat-ftubble, allowing it to remain until the middle of May following, that the fun might exhale the faline, oleaginous, and every other nutritive quality from it, which could enrich and fertilize the foil? It might well, as the writer obferves, be full of couch-grafs and other noxious weeds. He then, after ufing the couch-harrow, (which probably is an excellent inftrument) gave this land a deep earth, and fowed

fowed his buckwheat upon it; by which *deep ploughing*, he muft have buried and treafured up thoufands of feed-weeds, which could not have vegetated for want of fufficient tillage. Had he given the wheat-ftubble three fhallow ploughings, with good harrowings, allowing all the feed-weeds to vegetate between each ftirring of the earth, he might *then* with propriety have given it a *deep ploughing*, fpread his dung, and after turning it in, fown his feed with an almoft certain profpect of fecuring a good and profitable crop.

" The expences are doubtlefs charged as paid by your correfpondent; and he accounts in fome meafure for their magnitude, by faying he was *obliged to hire for every thing*. But to one who refides in a county where agriculture is perfectly well underftood and practifed, and buckwheat ufually cultivated, they appear too great to remain unnoticed. I fhall therefore contraft his expences for Inclofure No. I. with what the fame work would coft in Suffolk, fuppofing every thing put out by the day, viz.

	£.	s.	d.	Correfpondent's Expences.		
To ploughing 5¾ acres, at 4s. per acre	1	3	0	£.2	0	3
Four horfes and 2 men at dung cart, 2 days - - - - -	0	18	0	0	18	0
To ploughing 3 times in May, at 4s.	3	9	0	5	3	6
Dragging, a man and 4 horfes a day	0	8	0	0	8	0
	£.5	18	0	£.8	9	9

	£. s. d.	£. s. d.
Brought up - -	5 18 0	8 9 9
Couching and carting off - -	0 10 0	0 10 0
Six bushels seed fully sufficient, 5s.	1 10 0	2 5 0
Sowing and rolling - -	0 4 0	0 4 0
Rent - - - -	5 15 0	5 15 0
Tithe - - -	1 1 0	1 1 0
Poor-rates - - -	0 10 0	0 10 0
Mowing at 1s. per acre -	0 5 9	0 12 0
Turning, &c. by men and children, not only quite unnecessary, but detrimental - -	0 0 0	1 14 6
Threshing 21 sacks and 1 bushel, at 7d. per sack - -	0 12 4½	1 0 3
	£.16 6 1½	£.22 6 6

OBSERVATIONS.

Buckwheat is esteemed in Suffolk, Norfolk, and Essex, as adapted to lands of about 5s. value per acre; or from 3s. to 10s.; but can never answer on deep friable loam worth 1l. 1s. per acre; for on the *latter* there will always be too much straw, even without manure; and the *former* will yield on an average from three to four quarters per acre. It prevents the growth of weeds, but will not destroy couch-grass. It is usually sown in Suffolk with grass-seeds for laying down land, and for that purpose it is preferred to most other kinds of spring corn.

ARTICLE

Article XVII.

Description of the Construction and Use of a new Implement in Husbandry, for Transplanting Turnips.

[In a Letter to the Secretary.]

SIR, *Isle of Wight, Nov.* 18, 1784.

HAVING made a tour lately into Norfolk, among their many improvements in agriculture and its implements, shewn me, I was particularly pleased with their *Turnip Transplanter*. As a member of your society, and being willing to contribute what little assistance may be in my power towards general improvement in agriculture, I have taken the liberty of sending you one, made under my directions in this place.*

From the simplicity and cheapness of the instrument, and the very easy manner of using it, (two great recommendations in all implements in husbandry) I cannot but think it may become generally useful. As it frequently happens in turnip fields, that large spots fail, it is used for filling up those spots, from the adjoining parts of the same field. It may also be very useful in gardens, for transplanting plants of different kinds.

* A figure of it is given in the annexed Plate, No. I.

The

The method of uſing it is, to hold the long handle with the left hand, and the ſhort handle with the right drawn up; put the inſtrument over the plant that is to be taken up, and with your foot force it into the ground; then give it a twiſt round, and by drawing it gently up, the earth will adhere to the roots of the plant in a ſolid body; then with another inſtrument of the ſame ſize, take the earth out where the plant is to be put, and bringing the inſtrument with the plant in it, put it into the hole which has been made by the other; then keep your right hand ſteady, and draw up your left, and the earth and plant will be left in the hole with the roots undiſturbed.*

When turnips are to be tranſplanted in a field, there are two men employed with each an inſtrument, one man taking up a plant, while the other fills his inſtrument with earth only, thereby making room for depoſiting the plant; ſo that the hole which is made by taking up the plant, is filled with the earth taken out where the plant is to be put; which having depoſited, he takes up a plant, and returns to the place he firſt ſet out from, the firſt

* This inſtrument, which may be had at the ſociety's rooms, is well adapted for garden uſes, and particularly for gentlemen who would like the amuſement of tranſplanting ſmall roots of any kind, by an eaſy, clean, and expeditious method.

man at the ſame time returning with earth only; ſo that each man is alternately the planter, and each being employed both ways, the work goes on briſkly.

This inſtrument was invented by Mr. CUBITT GRAY, of *Southrepps* in Norfolk, a perſon who has given a great deal of attention to huſbandry, and particularly to the cultivation of turnips, for which crop he prepares his land in a different manner from moſt of his neighbours; they harrowing their land immediately after each ploughing, and then rolling it, in order (as they ſay) to keep in the moiſture; on the contrary, he never rolls his land, nor harrows it till he is going to plough it again, but leaves it as open as poſſible, in order to warm it, as he thinks land can never be too warm or dry for turnips; and he has always had the beſt crops, even when the ſeaſon has been dry when ſown. This method he has followed 16 years, and never once failed of a crop of turnips—though his neighbours frequently have. He has ſold turnips at five guineas and half per acre, to be fed off on his land: he always hand-hoes twice, as indeed do all the farmers in that country; his land is a ſandy loam, a very free working ſoil.

I am, your's, &c.

JOSEPH KIRKPATRICK.

[N. B. The Society return their thanks to Mr. KIRKPATRICK, both for his letter, and for the inftrument he fo accurately defcribes the ufe of. If he would favour them with an account of the mode of hufbandry practifed by the beft farmers in the Ifle of Wight, it would be efteemed an additional obligation.]

ARTICLE XVIII.

On the Cultivation of Broad Clover.

SIR, *Colfield, near Leith, Dec.* 9, 1786.

I AM duly favoured with your obliging letter of the 5th inft. and fhall be glad if the little that I have been able as yet to do, refpecting the culture of potatoes, fhould prove the means of ftimulating others, who have better opportunities, and greater abilities than myfelf, to attend, in a more particular manner than has yet been done, to the culture of this truly valuable plant.

It would give me a particular pleafure, if I could communicate to your fociety any thing that fhould deferve attention, refpecting the difeafe you take notice of, affecting Broad Clover. But as nothing *new* on the fubject has occurred in this part of the world, I cannot pretend to offer any conjectures as to the caufe, or hints for the remedy of that diforder. I have cultivated broad clover as a crop for more than thirty years paſt, and cannot ſay that I have here had occaſion to remark any thing *of late* that is in the leaſt particular.

During all my practice I have ever found, that although broad clover ſometimes affords as good a crop *the ſecond year* as the firſt, (obſerve, I call the firſt year of clover, that in which it firſt yields a crop, not that in which it is ſown) and on ſome occaſions even a better; yet I have ever found that that was in ſome meaſure caſual, and that no one could ſafely rely on it for a full crop the ſecond year. This is the caſe at preſent, as it ever has been in this part of the world.

There is no doubt, however, that broad clover is much leſs apt to fail in ſome ſoils than others. It is a plant that thrives beſt on a firm *weighty* ſoil. It therefore does very well in clays of a certain kind; (you will obſerve I make a diſtinction between *thriving* well, and *long life)* but on all clayey ſoils, and more particularly on ſoft ſpungy ſoils that have lately been brought into culture from moor, it is extremely liable to be thrown out by the ſeverities of the winter weather, and generally more ſo than on ſome others:—a firm hazel loam, or even a very *weighty*, or what we here call a *ſharp*, rich friable

mould,

mould, tending to a ſandy nature, is that on which I have ever found it leaſt liable to the accident of which you complain. I cannot tell if you have any of the ſoil of the kind I here deſcribe, never having been in Glouceſterſhire; and I am yet more doubtful if the terms I make uſe of, will be intelligible to you;—but I have no other means of communicating my ideas on this ſubject.

The proceſs by which broad clover is thus deſtroyed, is very eaſily obſervable on ſpungy moory ſoils, in which water is retained in a ſtate more nearly approaching to fluidity than in others. In rich clayey loams, where the ſame proceſs takes place in an inferior degree, the progreſs is far leſs perceptible. In ſuch ſpungy ſoils I have often remarked the following phenomena:—

After a night of bare froſt, in thoſe places where the earth is not covered with a cloſe ſward of graſs, the ſurface ſeems to be divided into a great number of broad kind of points, divided from each other by a great variety of fiſſures, ſomething like what takes place in a clayey pool, when the water has been ſuddenly evaporated, and the mud haſtily indurated. On taking up one of theſe detached pieces, and examining it, we diſcover that it conſiſts almoſt entirely of frozen water, with a thin cruſt of earth on its top. The ice in this caſe aſſumes a beautiful and ſingular form, conſiſting of a ſtack of needle-like columns, ſtanding in a vertical poſition, all of one height; a rude ſketch of which is given in the annexed plate A.* The vertical column conſiſts of tranſparent ſpiculæ, ſometimes with a little earth intermixed with them, but uſually pretty free from it. They have always a little earth at top; and when they are taken up, a little earth alſo adheres to their bottom; and below that the froſt has not penetrated. Theſe columns are longer or ſhorter according to circumſtances, from near two inches, as I have ſeen them, to leſs than a quarter of an inch. If the froſt continues more than one night, theſe icy pillars admit of a greater elongation by an additional range of columns ſhooting up below them, and forcing the former to riſe to a greater height. This ſecond ſeries of columns is always ſhorter than the firſt, and is divided from it by a thin ſtratum of earth, as at B.* Should the froſt continue longer, another row of ſhorter columns ſtill is formed beneath the former, divided from it alſo by its ſtratum of earth: and ſo it goes on, each night's froſt producing a new ſet of columns, which become gradually ſhorter, till at laſt the different ſtrata of earth which ſeparate them become ſo near to each other, as that the watery columns cannot be diſtinguiſhed;

* See plate No. I.

diſtinguiſhed; ſo that the baſe appears to be only a lump of frozen earth, to which theſe ſtacks of columns (C. D.) firmly adhere.

I ſhould not have taken the trouble to deſcribe ſo minutely as I have done this proceſs of nature, had I ever obſerved it deſcribed elſewhere; and becauſe it is very neceſſary to be adverted to, ſeeing it is uſually in conſequence of that proceſs that our broad clover fields are ſo frequently cut off by the ſeverities of our winter, which will be eaſily underſtood by attending to the following remarks.

When any vegetable is growing on the ſoil thus affected, the top of theſe columns of ice naturally lay hold of it as it were, and adhere to it ſo cloſely as to force it up along with them with a very ſtrong power. If the root penetrates the ground perpendicularly, either the main root muſt be broken off by this force, applied as at E, or the lateral fibres muſt be all torn off from the principal root, ſo as to leave it entirely naked, and thus expoſed to the injuries of the weather altogether defenceleſs. When a thaw comes, the columns of ice are melted away, and the earth ſubſides to its former bulk, ſo that the poor naked root ſtands up as at F, and muſt infallibly periſh. Every attentive farmer muſt have remarked his broad clover drawn at times out of the ground in this manner; though he has not

perhaps

perhaps adverted to the procefs above defcribed, by which it was effected. I have fometimes feen it thus forced out of the ground, on foils of the nature above defcribed, full fix inches; but in mellow cultivated fields, it feldom exceeds one or two inches.

Spungy foils, of the nature above defcribed, are therefore unfit for producing broad clover; and the nearer they approach to the nature of thefe, the more precarious will that crop be upon them: but nothing of this fort happens, when the froft is accompanied by fnow to a fufficient depth.

In foils of the nature here defcribed, it is obfervable, that if the furface be covered with fibrous-rooted graffes, they are not thus thrown out; for as thefe roots are generally numerous, and clofely interwoven with each other, they form fuch a clofe furface, that the tops of the icy columns do not appear divided; but the whole furface is lifted up with a confiderable ftratum of earth, fo that when the thaw comes, the whole furface fubfides together, and the roots quickly ftrike into the foil below; fo that though the furface might be eafily peeled off immediately after the thaw, in flakes, nearly as if it had been pared off by a fpade, yet in a fhort time the roots ftrike into the loofe earth below, and it becomes adhefive to the bottom as ufual. This phenomenon,

phenomenon, however, is never obfervable on any other foils but thofe that are uncommonly fpungy, particularly moffy foils; for the clofe furface of grafs prevents the froft from penetrating it in others: nor does the froft ever produce fuch devaftation on any foil, when well covered with a coat of matt-rooted graffes, as otherwife.

This being obvioufly the cafe, if ever reliance be to be had on broad clover for a *fecond* year's crop, it is advifeable to fow with the clover a proportion of rye-grafs. This is a very profitable practice, as it much augments the weight of the firft cutting, and makes it come fome weeks earlier than otherwife it would have been. It alfo effectually prevents the white gowans from appearing, which fo often render a crop of red clover fown alone worth very little. And as the clover afterward advances much more quickly than the rye-grafs, the fucceeding cuttings are as good as if no rye-grafs had been fown.

To guard the clover too, if reliance be to be had upon it for *the fecond year*, it fhould never be cut very late in the feafon, for this makes the furface fo bare as to leave the roots very much expofed to danger; but if it be cut pretty early in autumn, the rye-grafs advances again in the end of the feafon, after the clover has become ftationary, fo as to afford a clofe covering that defends the roots pretty well.

By

By theſe precautions, I have been ſeldom diſappointed in my ſecond year's crop of clover, though it will *ſometimes* diſappear almoſt entirely: nor do I think it poſſible in our climate ever to guard againſt this accident *with certainty*, where broad clover alone is ſown; ſo that I ſhould think it imprudent in any one, in almoſt any circumſtances, to rely on that *ſecond year's crop*; I have therefore ever held it as a maxim, that if a man is to depend on red clover alone, he never ſhould think of taking above one year's crop of it; but if he does reſolve to have two year's crops of graſs, he may always eaſily inſure that, and frequently have them of red clover, though not with certainty, if alone.

The rule I have ever followed to guard againſt every accident of this ſort, is, to ſow along with the red clover a conſiderable proportion of the white or Dutch clover, and ſome graſs. If the broad clover flouriſhes, theſe do not retard its growth, and only tend to thicken it; and if it ſhould fail, which it ſometimes will do in ſpite of every precaution, theſe plants fill the ground, and produce an abundant crop of herbage, which affords a greater weight and finer hay, than broad clover alone: though they do not anſwer quite ſo well for cutting for green forage.

White

White clover ſpreads its fibrous roots upon the ſurface of the ground, and is not ſo apt to be thrown out as red clover; nor is it ever deſtroyed by any accident, if the earth is rich and *firm*; frequent rolling makes it flouriſh abundantly, even on light ſoils; but without that, on ſuch ſoils it inevitably will periſh; (by the term *light*, I do not mean *ſandy*, as ſometimes is ſo expreſſed, but ground that is not *weighty*; we here call it *deaf.)*

If theſe haſty obſervations can be of any uſe to the members of your ſociety, they are much at their ſervice.

Some years ago, I publiſhed two volumes of "ESSAYS ON AGRICULTURE;" in which I threw out ſeveral obſervations on graſſes, but the above are not among them. There are few plants more valuable for *certain purpoſes* than broad clover. But there are many which *in ſome reſpects* exceed it. I muſt not however longer treſpaſs on your patience, than to aſſure you that I ſhall be ever happy to contribute my mite to the advancement of ſcience.

I am, your moſt humble ſervant,

JAMES ANDERSON.

N. B. The

N. B. The drawings [in Plate I.] are intended to give ſome idea of the progreſs of the *icy columns*. A. repreſents three ſtacks of one night old; B. ditto, of two nights' growth; C. ditto, of three; and D. a greater number, as they appear after the froſt has continued many days. On all occaſions there are numerous columns ſhooting up by the ſide of one another, ſo that the general ſurface of the ground is nearly as even as before the froſt; though that ſurface is by means of theſe columns lifted a good deal higher than its natural level. D. repreſents a few fibrous-rooted graſſes on the ſurface of the ground. At E. is repreſented a ſtalk of broad-clover, with its root forced up by the ſurrounding columns of ice, and broken off. F is the ſame root after a thaw, the columns of ice being melted, the earth ſubſided, and the plant falling down faded. A ſection of the earth is here ſuppoſed to be made to ſhew the portion of the roots.

ARTICLE XIX.

Biſhop of Killalue's Method of cultivating Potatoes.

[Given by him to Mr. WOODBINE.]

WHEN your land is left in ſuch an exhauſted condition, that it will not anſwer to plough it again for a crop, the method of bringing it again into heart by Potatoes is as follows;—

Spread your manure in lines (of about five or ſix feet broad) upon the ground, about twelve or fourteen

teen waggon-loads to an acre, leaving an interval of about two feet and a half between every row of manure. The intervals to be broader or narrower, according to the depth of ſoil on the land; where the vegetable mould is ſhalloweſt, the intervals to be broadeſt. Then cut your potatoes into pieces, leaving one eye (from whence a ſmall fibre of the root ſeems to grow) upon every piece: every one of theſe eyes will produce a new plant. Then ſpread the pieces on the ground, at a foot or a foot and a half diſtance. Then ſend in your diggers, and let them dig out of the intervals as much earth as will cover the pieces of potatoes about two inches.

As ſoon as the new plants all appear above ground, ſend in the diggers again, and cover the plants completely. When they appear above ground a ſecond time, cover them again with earth dug out of the intervals, taking care not to go much deeper than the remaining vegetable mould; though you may venture to go a little into the thill or clay, as it will tend rather to improve the land than otherwiſe; for lying at the top, it will not injure the vegetation; and being expoſed to the ſun and dews, it will be converted into fertile earth, in a ſeaſon or two, as well as the reſt.

When

When the weeds have appeared and are fit to pull, the crop muſt be carefully weeded, and in the courſe of the ſummer, muſt be weeded a ſecond time.

If the potatoes are planted in the latter end of March or even the beginning of April, they will be come to their full growth before Michaelmas. They muſt then be dug out, and the land will be left in condition to bear a good crop of wheat to be ſown at that ſeaſon with a ſlight ploughing, at which time the brows of the ridges ſhould be partly ploughed into the trenches, that the ground may be in order for future crops, and then the whole of the field properly covered with the crop. After this huſbandry, the ground will be fit for a crop of barley to ſucceed the wheat, and then a crop of oats with clover, &c.

N. B. By this courſe of huſbandry, the arable land of the farm will never be fallow for a year, as the potatoe crop ſucceeds the laſt crop of oats, and will be well worth 20l. per acre; and the land by the digging will be left in finer tilth than four ploughings will produce.

I twice tried an experiment, which anſwered beyond my expectations. Inſtead of firſt digging out my potatoes, I cut the haulm with a ſcythe, and threw it into the trenches. I then ſent the ſower to ſow the land with wheat; then I had the potatoes

dug

dug out, and let the wheat take its chance of being properly covered in the digging, and then gave it a slight harrowing; and by this method I had a crop of ten barrels to the acre; which I ascribed to the seed being better covered by being dug in, than it would have been by the harrow in the usual way. I do not, however, recommend this to be done the first time this husbandry is tried, though the experiment may be made in one ridge only, and according as that succeeds, it may be pursued or not hereafter.

July, 1786. THO. KILLALUE.

Article XX.

Observations on the Disease called the Wind in Sheep.

[By Mr. J. Webb, Apothecary.]

Gentlemen, *Doynton, Gloucestershire.*

NUMEROUS as the diseases of the brute creation are, I believe they suffer but little less from them than from the absurd means that frequently are administered for their relief, arising from the generality of farmers being very ignorant both of the *seat* and *cause* of the complaint. This

I attri-

I attribute partly to the terms ufed for difeafes not conveying any juft and proper ideas of them.

I have feen feveral fheep, immediately after being fhorn, appear to be in violent pain; their fides are fomewhat extended, and their breathing very fhort; the head is hung drooping, and they have a great averfion to moving or walking, and generally lie down. Thefe fymptoms continue increafing till the fheep dies in a few hours, unlefs a violent purging come on, which generally gives immediate relief. On enquiring for the name given to this complaint, I found it was called *the Wind*; but where the feat of it lay, few could tell. Some thought it was in the head, others in the lungs or lights, &c.; and the remedies they applied were as various as their opinions of the difeafe; fome giving gin, others black pepper, or both thefe mixed together. Daffy's elixir, and elder-berry fyrup, are fometimes administered.

Not fatisfied with thefe accounts, I endeavoured (by infpecting the carcafes of fheep that died of the difeafe) to difcover the caufe and feat of the complaint. On opening four fheep that died of the difeafe, I found all the inteftines rather diftended with flatus,* but not in any great degree. Their

* From whence I fuppofe the term *wind* for this diforder originated.

blood-

blood-vessels were very turgid and of a deep red, particularly those of the large intestines, excepting the rectum, (or what is commonly called the *bum-gut)* which had a healthy appearance, as likewise had the stomach, milt, caul, liver, heart, lungs or lights, &c.; and in short all the viscera contained in the cavity of the trunk. From these appearances I will venture to say, that the disease in question is a violent inflammation of the intestines; perhaps in some measure arising from bruises in shearing, but more so from losing a warm clothing, and being suddenly exposed to cold air and cold feeding.

I beg leave therefore to recommend to farmers, that on the first appearance of the complaint, they put the sheep into a stable or other warm place, and immediately bleed it very freely. Bruise a quarter of an ounce of some carminative seed, such as carraway, anise, cummin, or fennel, and mix these with two ounces of Glauber's purging salts in a pint of water; place it on a fire, and make it boil for a few minutes, then strain it off: then add a quarter of an ounce of powdered jalap, and while lukewarm give the sheep a quarter of a pint of this liquor (first well shaken together) every half hour till it dungs. It should have no food or cold water till recovered, but a little warm water might be of service.

 This

This remedy, I imagine, might be of ſervice to oxen, when blaſted from putting them into freſh clover; but they being much larger and ſtronger animals, will require a doſe larger in proportion. With the ſame regulation I would recommend it for the fret in horſes, as a better remedy than the drenches commonly given.

I am, Gentlemen, your's &c.

Oct. 10, 1786. J. WEBB.

P. S. Perhaps it may appear ſtrange to recommend gliſters for horſes; but I am well aſſured that one (compoſed of ſome tobacco boiled in a pint of water a few minutes, then ſtrained off, and with the addition of a little ſweet oil or hog's lard) given as ſuch, would greatly aſſiſt the drench before recommended.

Article XXI.

On the Cultivation of Apple-Trees.

SIR,

I Have ever eſteemed Apples as the moſt uſeful fruit cultivated in this kingdom. They are placed on the tables of the great, and are within the power of the cottager to enjoy; at whoſe homely board, when dreſſed in puddings or pyes, they may be conſidered as a luxury.

These

These considerations make me view with concern the present neglect of orchards, where the old trees are decaying without proper provision being made for the succeeding age: for if a farmer plants fresh trees, (which does not frequently happen) there is seldom any care taken to propagate the better sorts, as his grafts are usually taken promiscuously from any ordinary kind, most easily procured in his neighbourhood.* Hence arise the numberless apple-trees, wich may almost be said to incumber the ground, and occupy the room which a valuable tree might possess.

I have heard it frequently remarked, that a good apple is hardly ever to be procured but near large towns; and in general I have found the observation just, owing, I conceive, principally to the inattention of the farmer, and sometimes to the difficulty he finds in obtaining the best sorts.

* We hope for the credit of common sense, this is not literally the case; few people would take the trouble of grafting without a view to some improvement; and in country places, the best apple-trees of the neighbourhood are sufficiently noted; but if the spirit of improvement prevailed as it ought in this article, the owners of orchards would not only propagate the best sorts *occasionally*, but constantly reduce their young apple-trees which were found to bear ordinary fruit, to grafting stocks for the most valuable sorts. By such a practice, great reformation would be made.

Could these difficulties be obviated, I conceive we should in a few years find good apples at every cottage, and greater choice at our country markets.

This being, in my opinion, an object of some utility, give me leave to suggest some hints, which you may possibly improve and render really useful.

After making a catalogue of the best apples for eating, baking, keeping, &c. suppose your society were to procure grafts of each kind from the counties most celebrated for the respective sorts; which, I imagine, may be done by a society at a moderate expence, as many gentlemen would be proud to furnish them; so that the carriage would be the only cost: some might be purchased and given away, without any great expence to the society.

I fear offering a premium for the cultivation of apple-trees, would be inadequate to the intent; as small rewards, which must necessarily be difficult and slow in their operation, would not raise a spirit in farmers in general to pursue the object.

I therefore think that the Bath Society (if they deem the subject worthy notice) might advertise that they were collecting a quantity of grafts from apple-trees of superior kinds, which they meant to

distribute

diſtribute gratis to farmers and cottagers who applied and engaged to cultivate them; and that the application muſt be made either verbally or in writing by a certain day, ſpecifying the number the party wiſhes to have. And when the grafts are ready, a ſecond advertiſement might give notice for each perſon who has applied to call or ſend for the proportion allotted to him.

If the ſociety alſo recommended to their correſpondents and other gentlemen, a ſimilar diſpoſal of good grafts, round their reſpective neighbourhoods, I ſhould not doubt but in a very few years every county would be plentifully ſupplied with the beſt apples.

I am, Sir, your's, &c,

RICHARD SAMUEL.

[N. B. The ſubſtance of this letter is of great importance to the nation; for it muſt be univerſally acknowleged that the apple is the firſt of fruits, as an article of family conſumption, if not as a luxury. In its different applications for cyder, for dreſſing as food, and for eating in its natural ſtate, its flavour is of great conſequence to our pleaſure, and perhaps of ſome to our health. For it is not eaſily ſuppoſable, that a rough acrid cyder is equally wholeſome with a ſoft and pleaſant ſort, to all conſtitutions;—and while the ſame ſpace of ground, and the ſame expence, will ſuffice for a fine ſort of apple-trees, as for one of the worſt, it is well worth publick attention to promote the growth of the fineſt ſorts, as at once cheap, wholeſome,

 and

and grateful. It is a *benevolent* object, likewife, when we confider how eafily the palates of the middling and lower claffes of our countrymen, who cannot afford the luxury of more expenfive fruit, may be gratified by improving the quality of this common and excellent fpecies.

From thefe confiderations, it is hereby requefted, as a firft ftep towards general improvement, that thofe gentlemen, who have been particularly curious in the improvement of their orchards, or in remarking the diftinctions and excellencies of different forts of apples and pears, would be fo obliging as to communicate their obfervations on the fubject by letter to the fecretary, as foon as they conveniently can after reading this article. They are earneftly requefted alfo to mention particularly the flavour and properties of the forts they approve beft, for fummer ufe and for long keeping; likewife with what number of fcions they could conveniently furnifh the fecretary for diftribution; firft among the members of the fociety, and fecondly to others who may wifh to concur in the general plan of improvement.]

ARTICLE XXII.

On the Degeneracy of Apples.

[In a Letter to the SECRETARY.]

SIR, *Kenfington, Dec.* 16, 1786.

THE fubject of the decay of the different forts of Apples, I have maturely confidered, and compared my ideas with thofe of men in long practice with myfelf, and find it is the general opinion, that

that it is not a real decline in the quality of the fruit, but in the tree, owing either to want of health, the ſeaſon, the ſoil, the mode of planting, or to the ſtock which they are grafted on, being too often raiſed from the ſeed of apples in the ſame place or county: it appears from the ableſt men in my profeſſion, that they never found a real decline in any one kind of fruit, but from the above cauſes.

To make a fair experiment, I ſhould be much obliged to any gentleman that will take the trouble to ſend me a few cuttings, from thoſe very trees, the fruit of which is ſuppoſed to be degenerated from the original goodneſs. I would graft them on the real crab-ſtock, and ſome alſo on the ſtock raiſed from the apple-pips in this county; then ſend the trees to the place where the cuttings came from;— by which means (though the proceſs is tedious) we ſhall be able to aſcertain, whether the change of ſtock will not reſtore the fruit to its original goodneſs.

I have not a doubt in my own mind, but that the trees which are grafted on the ſtocks raiſed from the apple-pips, are more tender than thoſe grafted on the real crab-ſtock; and the ſeaſons in this country have for many years paſt been unfavourable for fruits, which add much to the ſuppoſed degeneracy of the apple. It is my opinion, that if planters of orchards would procure the trees grafted on real crab-ſtocks from a diſtant county, they would find their account in ſo doing much over-balance the extra expence of charge and carriage.

My reaſon for recommending the true crab-ſtock is, that I believe the crab to be a native of this country; but whether it is or not, we are ſure it is much hardier than the ſtocks raiſed from apple-pips, and there cannot be a doubt but the apple was originally an exotic. The crab-ſtock will ſucceed in many different ſoils, particularly in ſtiff, cold, moiſt ground, where the apple-ſtock will canker and die. I recommend the crab-ſtock for this reaſon, as alſo that it is not ſo early in vegetating as the apple; by which a few days may preſerve the flower from the cold blaſts, and be the means of ſaving a fine crop of fruit.

Your moſt obedient ſervant,

DAN. GRIMWOOD.

Article XXIII.

On the Culture of Parſnips.

TO cultivate this root ſo as to make it prove advantageous to the farmer, it will be right to ſow the ſeed in the autumn, immediately after it is is ripe, or come to perfection; by which means the plants will appear early the following ſpring, and will get ſtrong before the weeds can grow to injure them. Froſts never affect the ſeed, nor do the young plants ever materially ſuffer through the ſeverity of the ſeaſons. Not only on this account, but for many other reaſons, the autumn is preferable to the ſpring ſowing, as the weeds at this time will keep pace with the parſnips; and often when they are hoed or cleaned, great part of the crop is pulled up, cut out, or otherwiſe deſtroyed, as they are (when ſown in the ſpring) ſo ſmall when they firſt appear, as not eaſily to be diſtinguiſhed from the weeds; and if no rains fall at that ſeaſon, ſome of the ſeed will not vegetate till late in the ſummer; and the few plants that do appear, will ſcarce pay the expence of cleaning them; beſides, they will never grow to any ſize, but be ſticky or cankered, and conſequently will be deſtitute of nutrimental juice; while on the contrary, thoſe that are ſown in the autumn will be large, free from the defects of the others, and fully anſwer the expectation of the cultivator.

The beſt ſoil for parſnips is, a rich deep loam; next to this is ſand, or they will thrive well in a black gritty ſoil; but will never pay for cultivating in ſtone-braſh, gravel, or clay ſoils; and they always

are

are the largest where the earth is the deepest. Dry light land is pleasing to them, but wet, stiff, or hide-bound land is destructive. If the soil be proper, they do not require much manure. The writer hath obtained a very good crop for three successive years, from the same land, without using any; but when he laid at the rate of about forty cart-loads of sand per acre upon a very stiff loam, and ploughed it in, he found it answered very well, from which he concludes that a mixture of soils may be proper for this root.

It is most adviseable to sow the seed in drills at about 18 inches distant from each other, that the plants may be the more conveniently hand or horse-hoed; and they will be more luxuriant if they undergo a second hoeing, and are carefully earthed so as not to cover the leaves.

Those who have not ground to spare, or cannot get it in proper condition to receive the seed in the autumn, may at that time sow a plot in their garden, or the corner of some field, and may transplant from thence the latter end of the month of April, or early in the May following. The plants must be carefully drawn from the seed plot, and the land that is to receive them should be well pulverized by harrowing and rolling; and when it is thus ordered, a

furrow

furrow ſhould be opened with the plough about ſix or eight inches deep, in which the plants ſhould be regularly laid at about the diſtance of ten inches from each other, taking care not to let the root be bent, but for the plant to ſtand perpendicular after the earth is cloſed about it, which ſhould be immediately done by means of perſons who ſhould for this purpoſe follow the planter with a hoe; and he muſt not forget that the plants will be injured if the leaves are covered. Another furrow muſt be opened about 18 inches from the laſt, in the ſame direction, and planted as before; and ſo in like manner till all the plants are depoſited, or the field is completely cropped; and when the weeds appear, hoeing will be neceſſary, and it will be right afterwards to earth them.

There is no doubt but many may diſapprove of the method of tranſplanting parſnips, yet ſome may be induced to try the experiment, when they conſider that they may perform it at a time when there is little beſide to be done in a farm, and that their crop will be more certain; for if they are planted after rain, they will not be checked by the removal, nor will they be injured by weeds, or the ground ſuffer ſo much by being thus planted, as otherwiſe it would do, if the ſeed were ſown in the autumn.

It is wrong to plant parſnips by means of dibbling, as the ground thereby becomes ſo bound as ſeldom to admit the ſmall lateral fibres (with which theſe plants abound) to fix or work in the earth, on which account they are prevented from expanding themſelves, and never attain their proper ſize.

If people would in general be attentive to the ſoil, the ſeaſon for ſowing, the cleaning and earthing the plants, and raiſing their ſeed from the largeſt and beſt parſnips, (which ſhould be ſelected and tranſplanted for this purpoſe) there is no doubt but ſuch a crop would anſwer much better than a crop of carrots; they are equal if not ſuperior for fatting pigs, as they make their fleſh whiter, and they eat them with more ſatisfaction. When they are clean waſhed and ſliced among bran, horſes eat them greedily and thrive therewith; nor do they heat them, or like corn fill them with diſorders.

It is reported, that cows and oxen are fond of parſnips; if ſo, they are certainly well worth a farmer's attention, eſpecially in countries where there is a ſcarcity of fodder. The writer therefore flatters himſelf, the foregoing directions may prove ſerviceable to gentlemen or farmers, who are ſo circumſtanced; and begs to aſſure the Society he ſhall

all times be happy to communicate to them the little knowledge he possesses or may acquire.

Littleton-House,
June 1787.

J. HAZARD.

ARTICLE XXIV.

Account of the Culture, Expences, and Produce of Potatoes, per acre, about Ilford, and the adjacent Parishes, six miles East of London.

THE soil on which the best crops are raised, is a strongish loam, not quite approaching to clay. The manure used is rotten dung, which is laid on just before planting, in the proportion of from 15 to 20 loads per acre.

Twenty-four bushels, cut into sets of one or two good eyes each, are planted per acre, at 15 inches distance, and kept clean by hoeing, in which the earth is drawn up round the plants as they advance in height. The produce on an average, ten tons per acre—126lbs. to the hundred weight.

They take them up with a broad three-tin'd fork, at three guineas per acre. The average expence per

per acre, every charge included, is about ten pounds. Their value for feeding hogs, 25s. per ton; at which valuation they are worth only 12l. 10s. which leaves a profit of only 2l. 10s. But the growers generally make double that price at the London market, which brings the profit to 15l. per acre.—They seldom sow potatoes more than two years on the same land, without an intervening crop.

The *Aylesbury white* is mostly planted for the table; but the *Ox-noble* is most productive for cattle.

W. BRAINES.

Article XXV.

On the Depravation of Apple-Trees.

[From Mr. Gillingwater, of *Harleston*, addressed to Mr. John Wagstaffe, *Norwich*.]

I Observed in the Ipswich Journal of Saturday last, that circular letters were sent from the Secretary of the Bath Agriculture Society, relative to a representation made to that Society, viz. "That in most of the counties, and particularly in that of Worcester, the old and best kinds of apples are nearly lost; and that by persons conversant in fruit-trees it is apprehended, they will in a few years be entirely gone." I must acknowledge, that I was struck with the representation; and it immediately occurred to me, the conversation which we lately had at Harleston, when I observed to you the danger which orchards of apple-trees, when planted too near each other, were exposed to, from the mixture of various farina: and this, I apprehend, is the cause of the degeneracy of all the old and best kind of apple-trees in the great cyder counties of this kingdom, which is here complained of; and which the Society so earnestly requests its members to extend their enquiries concerning.

This conjecture appears to me extremely reasonable; for if the great variety of apples, and also other fruits, be produced by the casual intermixture of different farina, the fruit also itself must be affected. The old and best kinds of apple-trees, I apprehend, are not lost at all, but are only corrupted from being planted too near bad neighbours:—remove them to a situation where they are not exposed to this inconvenience, and they will immediately recover their original excellency.

The Society informs us, that their trees of the best kind are *nearly* lost, but not *altogether* so. The reason I conceive for this distinction is, that some few,

few, having the advantage of being situated where they are not injured by the farina of other trees of inferior kind, retain their primitive qualities; whilst others, which are planted indiscriminately in large quantities in orchards, are almost totally spoiled, from the farina of other surrounding trees, which intermixes with them.

Your's, &c.

Harleston, EDM. GILLINGWATER.
July 10, 1786.

An ADVERTENCE *to the foregoing.*

THE remarks of my friend respecting the probable alteration in the distinguishing quality and flavour of fruits, by an indiscriminate planting of various species of apple-trees together, are unquestionably well founded; but whether the entire depravity can be conquered, and a perfect regeneration of the original specifick quality of the fruit be recovered, is a matter of question. For we may consider the circulation of the sap in trees as somewhat analogous to that of the fluids in animated bodies; and that the latter imbibes salubrity and contagion from the approximation of different subjects; whereby a constitutional change is sometimes effected.

Now

Now the degeneracy of the beſt ſpecies of apples from the aforementioned cauſe being allowed; and as there is an acknowledged flow of the elementary fluid to the bloſſom, and to its fruit, and in refluent ſucceſſion from both, what theſe receive or imbibe may, by a repeated circulation, alter the habit of the tree. This ſpeculation might have been too much refined, had not it been experienced, that a ſcion ingrafted hath not always produced that ſpecifick fruit from whence it was preſumed to be taken; and that the mere inſertion of the bud in inoculation hath, without ſenſible vegetation, altered the habit of the plant in which it was inſerted.

The firſt inſtance has been atteſted by ſome practical obſervers; the latter is founded on an experiment related by BRADLEY, under his "particular proofs of the ſap's circulation in plants;" where he mentions the "inoculation of ſome of the *paſſion-tree*, whoſe leaves were ſpotted with yellow, into one of that ſort of paſſion-tree which bears the long fruit. Now though the buds did not take, yet in a fortnight's time the yellow ſpots began to ſhew themſelves above the inoculation, and in a ſhort time after appeared on a ſhoot which came out of the ground from another part of the plant."

Neverthelefs, the difcriminate planting propofed by my friend will generally apply to the prefervation of the original diftinction of the fruit; for whether the farina be wafted by the breeze, or winnowed by the wings of infects, it muft be in a contiguity of trees that the alteration muft arife.

It is true, that bees, wild and domefticated, with many other infects, infert their bodies within the *nectarium* of the bloffom, and that there is a frequent adhefion to their downy fides of the impregnating duft; which is not unfrequently conveyed to a various fpecies of bloffom, from that whence it was received.—Still, were it not for the contiguity of the various trees, no fenfible change would be effected by the infect becoming an *auxiliary* to the *furer* impregnation of thofe trees of the genus wifhed to be preferved from depravity. This precaution will equally apply to a valuable fpecies obtained from feed, or an undegenerated old fpecies to be extended; as the planting *either*, fomewhat remote from other apple-trees, will be certainly out of the flight of the farina, or the probable conveyance of it by infects; which rarely quit a vicinity that fupplies their nutriment, but to depofit their extract in the common repofitory.

It is a piece of juftice to advert to fome fubfequent remarks from my friend, that "no degeneracy

neracy is to be apprehended from the proximity of other fruit-trees; as the admirable disposition of the receptacle to its farina, denies every ungenerical impregnation."

Norwich, JOHN WAGSTAFFE.
Nov. 9, 1787.

Article XXVI.

On the Cultivation of Flax and Hemp.

SIR, *Wisbech, July 11, 1784.*

IN answer to your letter, I have applied to a friend of mine who has had many years' experience in cultivating both Flax and Hemp; and he informs me as follows, viz.

1st. The usual price of flax-seed is 2l. 2s. per coomb; the quantity sown is two bushels per acre.

2dly. As to manure, there is none laid upon land where you intend to sow flax; but it must be clean or sward land.

3dly. If the land be sward land, or what we call grass land, it must be ploughed but once, and harrowed fine. As to my rich land, it will bring turnips, wheat, or clover, &c. after the flax is off.

4thly. The crop, managings and getting into the barn, costs somewhat about 3l. per acre. The produce is from 20 to 50 stone per acre, according to the land. As to the score, I have sold at 5l. 6l. and 7l. per score.

5thly. It does not impoverish the land, but rather improves it.

P.S. You should have an experienced man to sow it, as there are but few who understand that business. And likewise when it is fit to pull, one who knows when to put it into the dike, and when to take it out, as there is a great deal of difficulty in managing that point, without spoiling the flax.

I am, your humble servant,

JAMES ELLERKER.

ARTICLE XXVII.

Description of a Comb-Pot, to be used with Pit Coal.

Invented by JOHN ASHMAN, of *Abbey-Milton, Dorset.*

SIR, *Sarum, Oct. 24, 1786.*

THE sketch of a comb-pot I here send you, was invented by JOHN ASHMAN, who has been in our service about six months. He worked

it

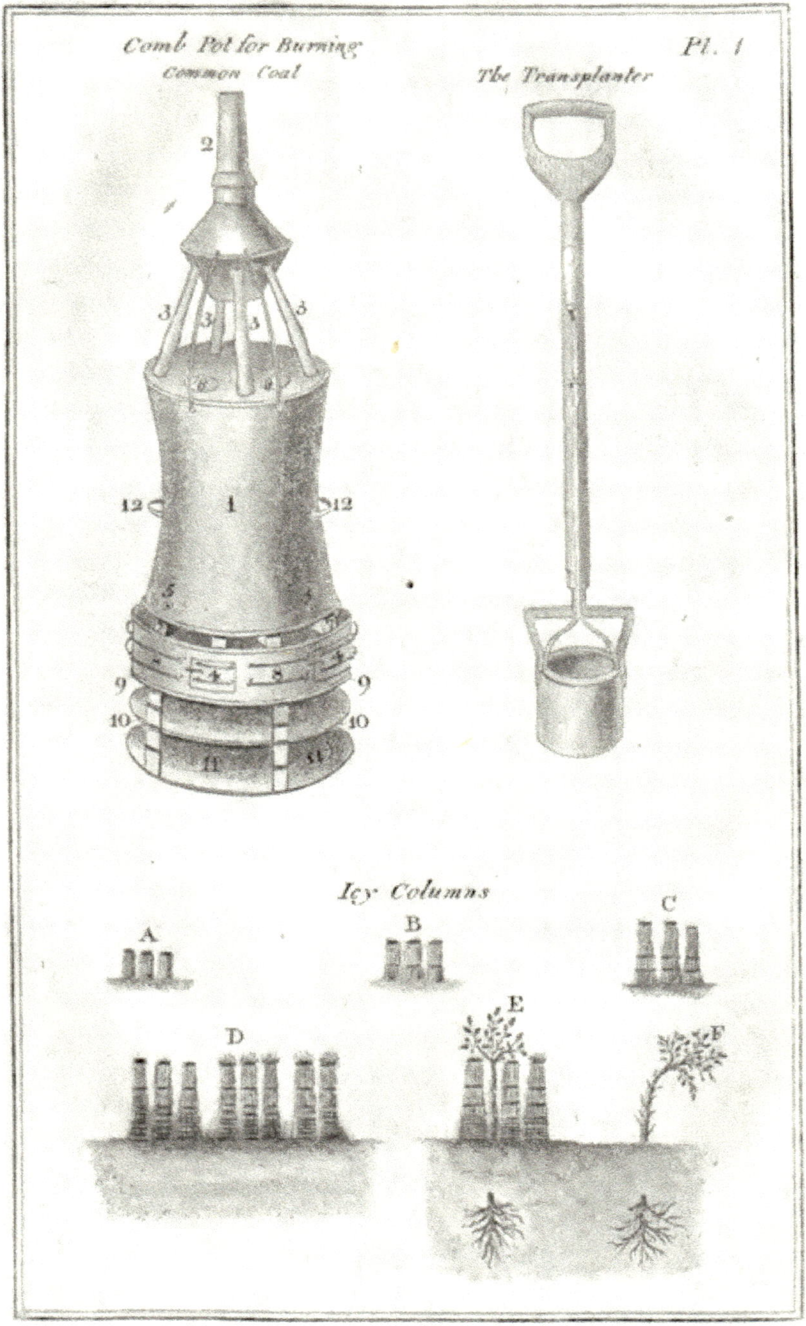

Comb Pot for Burning Common Coal

Pl. 1

The Transplanter

2
3 3 3 3
6 6
12 1 12
5 5
4 8 4
9 9
10 10
11 11

Icy Columns

A B C
D E F

it two years and a half at *Abbey-Milton*, and three years and a half at *Blandford*, and is very anxious to present it for the approbation of the Bath Society.

Your obedient servant,

DAN. & THO. DYKE.

PLATE I. *Fig.* 1. The furnace for water, which contains a smaller one, keeping the suds of the second washing the wool, for to be used with the next quantity of wool the first way.

Fig. 2. A tin chimney for conveying the smoke, (and carried higher in any direction made of tin) the lower part of which is made almost globular, for the better conveniency of taking away the four smaller ones from the top of the furnace, [*fig.* 3. 3. 3. 3.] to clean the same tubes continued through the furnace, close to the side at equal distances, and directly over each fire-place.

Fig. 4. 4. (with two more on the other side) The fire-place doors.

Fig. 5. 5. Cocks for drawing off the water and suds.

Fig. 6. 6. Covers to the furnaces.

Fig. 7. 7. (with two more on the other side) Spaces between each fire-place, for receiving the combs to heat on a cast-iron plate.

Fig. 8. 8. (and two more) Large wires on iron-plates, projecting a distance sufficient to prevent the wool from singing in the combs while heating; each place heating one pair of combs.

Fig. 9. 9. An iron plate, for making the fire on, with holes to let the ashes through.

Fig. 10. A stone to receive the ashes on; and at four equal distances bricks to support the upper part.

Fig. 11. A place for the pit-coal, supporting in like manner the ashes. Plate, &c. as the last described.

Fig. 12. 12. Handles for taking off the furnace.

The above pot is made of a circular form, lessened in the middle for receiving the handles of the combs while heating.

Article XXVIII.

On the Disorder called the Blast in Sheep.

Sir, *Wiley, Wilts, May 7, 1787.*

IN our county we breed many sheep, and manure the fallows, &c. with them. After having penn'd them all night, when they are driven into fresh grass, or young clover, they are frequently taken with what we call the Blast; that is, they over-gorge themselves, foam at the mouth, swell exceedingly, breathe very quick and short, then jump up, and fall down dead instantly. This is so frequent a disorder, and so great a loss, that a neighbour of mine had 17 die in one morning—indeed within half an hour; for they are often taken with it many at a time. We have no remedy, ever known as yet, but

but driving them into a bare place like a road, and keeping them in motion;—but it is so sudden, there is not time for that in general. It is a disorder not unfrequent in cattle; and having a cow taken in this manner, I had heard that, by stabbing her in the maw, I stood a chance of saving her life—I did this; the matter flew out, gave instant relief to the cow, she did well, and has had two calves since. I therefore resolved to try the same with my sheep, and have succeeded to my wish. The way I perform it is as follows:—

The sheep will swell considerably on the left side (or what you would call the nigh side of a horse) near the kidneys, behind the ribs, which is near the flank; the swelling is very protuberant, and there is mark enough, (about three inches) where if you dart your knife in, you must *at this time* go instantly into the maw; the food or matter immediately flies out, gives relief, and with only common applications of a horse-doctor's mixture of bees-wax, rosin, grease, &c. the sheep is sure to do well. All my neighbours were surprised at my success, as it was quite new to them and to all the shepherds around.

I am, Sir, your humble servant,

WM. POTTICARY.

ARTICLE

Article XXIX.

A Continuation of Experiments in the Drill Hufbandry.

[By Sir John Anstruther, Bart.]

Experiment of Drilled Barley—Crop, 1786.

IN ——— 1785, 3 A. 2 R. 20 P. of Englifh meafure were drilled with fome Lincolnfhire barley. Part of this was one acre, upon which the experiment of tranfplanted wheat, and dibbled wheat and barley, was made laft year. The remainder was after carrots and cabbages, and as it was not manured for thefe, it had a top-dreffing of dung. Upon that were drilled 4½ bufhels of Winchefter meafure; thefe were drilled after the plough, by a man following the plough, and dropping the feed by hand; a fecond plough followed and covered this; and the third furrow was fown and covered in the fame manner: by this plan the rows were at 18 inches diftance; thefe were hand-hoed once at the expence of 10s. The produce was 147 bufhels, which is nearly 32 for one, and is 40¾ bufhels per Englifh acre.

The reft of the field was fown broadcaft at the fame time, with the fame grain; the fame quantity of this was meafured, viz. 3 A. 2 R. 20 P. and the proportion

proportion of feed to this meafure was 16½ bufhels, or about 4¼ per acre. This was after turnips, for which the land was previoufly manured. The produce was 114 bufhels, which is nearly 7 after one, and 31½ bufhels per Englifh acre.

	Bushels.		Bushels.
Produce of the drilled per acre - -	40½	Broadcaft produce	31½
Deduct the feed -	1¼	Deduct the feed -	4¼
Clear produce	39¼	Clear produce	27¼

Produce of the drilled fuperior 12¼ bufhels.

In the account of the experiment of the drilled barley laft year, it was omitted to mention the quantity of ground drilled, and of the broadcaft. The quantity was 2 A. 20 P.; and the fame quantity was meafured of the field which was in broadcaft:—it was drilled as above defcribed, and fown under furrow.

Experiments 1786.

Half an acre was drilled with wheat, and horfehoed. This was once ploughed after a crop of barley, drilled at 18 inches. It was defigned to have been ploughed in ridges 4 feet 8½ inches, which fhould have made 21 ridges, as the breadth was 33 yards; but by the ploughman's want of experience in ploughing ftraight ridges, they were unequally broad, and there were only 17 ridges.

It

It was drilled the 21ſt of October 1785, with two rows of wheat upon each ridge, with partitions of 14 inches; two intervals three feet four inches. It was ſowed with a peck and a half; was four times horſe-hoed; twice from, and twice to, the rows; and three times hand-hoed in the partitions and rows.

October 21ſt, upon a ridge 74 yards long, and four broad, which is the 16th part of an acre, were tranſplanted ten rows of wheat plants, at nine inches diſtance every way, from ſeed ſown in a garden the end of Auguſt, and two rows from ſeed ſown in May.

Nov. 12th, another ridge of the ſame dimenſions was planted with plants (from ſeed ſown in Auguſt) at the ſame diſtance. At the ſame time one ridge was dibbled of the ſame dimenſions with wheat, at nine inches, and the wheat dropped in the holes, and from one grain to 15 per hole, and two rows of each.

March 31ſt, two ridges of the ſame dimenſions (viz. the eighth of an acre) were planted with plants, at the ſame diſtance, from ſeed ſown in Auguſt.

In July and Auguſt, viewed the experiments, and found a high wind ſome days before had broken down a great deal of the drilled wheat; and that, in the horſe-hoed, the earth had not been properly ploughed up to the rows, and as the earth was fine and looſe, it had ſo fallen down from the roots that the plants had little ſupport from the earth on one ſide, and the weight of ears with the high winds had made them fall over, by reaſon of that want of ſupport. The earth was hoed up to the rows, which I found ſupported the ſtems from falling over. The very dry ſeaſon, or the froſt in winter, or ſome other cauſe, had deſtroyed a great deal of the drilled wheat, as well as the dibbled and tranſplanted. In many places there was from one to two feet deſtroyed: theſe had been filled up by tranſplanting the 21ſt of April, but many of them died; or were ſmall plants and ſmall ears, and not above three or four ears to a plant.

There appeared at this time no difference between what was tranſplanted in October and November, or between thoſe from the ſeed in May and Auguſt. But the two ridges tranſplanted the 31ſt of March was the worſt crop, and much greener than that tranſplanted before winter, having but nine or ten ſtems on the beſt plants; but

many

many of the other had 16. The ridges dibbled with grain in November, appeared at this time a better crop, and the ears larger than the tranſplanted. On counting the ears, it did not appear there were more from the holes where there were 12 and 15 grains, than where there were fewer grains planted.

The tranſplanted and dibbled were much leſs layed over by the wind than the drilled; which appeared to be from the ſtems ſupporting each other, and the earth about the roots being firmer than the drilled, which had leſs ſupport on one ſide, from the earth not being properly layed to the ſtems. Theſe were reaped the 22d of September. That tranſplanted in March was not quite ſo ripe as the other.

The drilled half acre produced five buſhels and two pecks; which is at the rate of 11 buſhels per acre:—had there been 21 ridges, it would have been in proportion 13 upon 21 ridges. The numerous gaps or blanks, and ears broke down and loſt by the wind, made this a bad crop.

To ſee what the produce might have been if it had been equally good, the ears and grains of ſome yards were counted of the drilled rows, and, where equally

equally good, ſome yards of a ſingle row produced 124 ears, ſome 102; and the average of theſe counted, was 113 ears per yard of a ſingle row. Some ears produced 80 grains per ear; the loweſt was 50, and the average 61; and the number of grains per ounce was 880. The ears multiplied by the grains, and divided by 880, is $7\frac{7}{8}$ ounces per yard, of ſingle rows; there being 14 ridges of the above breadth in an acre, this made 5992 yards of ſingle rows, excluſive of head ridges; this would be at the rate of 49 buſhels per acre, had it been all equally good, and 28 rows in the breadth of the acre.

	B.	*P.*
The produce of the two ridges tranſplanted in October and November, being the eighth of an acre - - - -	2	2
That of the two tranſplanted the 31ſt of March -	1	0
The one ridge dibbled with wheat in different quantities, being the 16th of an acre - -	1	0

At this rate the proportions would be per acre,

That tranſplanted before winter - - -	20	0
That planted with grain - - - -	16	0
That tranſplanted the 31ſt of March - -	8	0

The produce of theſe experiments is but ſmall; but could they have been compared with the broadcaſt this year, they would have made a better

appearance

appearance than by comparing them with good crops of other years, as the broad-caſt crops of wheat in general were very thin and bad this year. One thing was to the diſadvantage of theſe experiments:—they were made upon a part of that which was in drilled barley laſt crop; and ſome of the barley had been ſhaked out, which ſprung up early, and made it neceſſary to hoe it early to deſtroy it. Of the tranſplanted, a great many of the plants had 16 ears; and if a crop were equally good, the produce would be very great, even allowing each to produce on an average eight ears, as each plant has a ſpace of nine inches ſquare, this is 77440 plants, and at the above average of the drilled at 61 grains per ear, and 880 grains per ounce, the produce would be 44 buſhels; therefore it appears the ears were not above four per plant.

The ſmall produce of the tranſplanted, it may be ſuppoſed, might have been occaſioned by being badly planted; but the dibbled was worſe, that was not liable to that accident. And we do not find the planted wheat, where practiſed, produced ſo great crops as might be expected.

Whether theſe methods are more liable to accidents than the broadcaſt, experience and more extenſive practice may diſcover.

These small unsuccessful experiments by no means prove the method bad, but the bad execution or unfavourable season; and from these we see what great produce they may yield when properly executed in more favourable seasons; as in a former experiment the drilled was much more successful, although this is so bad. And it is to be hoped, as many of the intelligent correspondents of the Society have practised the drilled husbandry, that their experience will shew it to be as profitable as many former practisers of it have shewn.

J. ANSTRUTHER.

Article XXX.

An Account of a Series of Experiments made by Mr. Nehemiah Bartley, *on his Farm near Bristol.*

[In a Letter to the Secretary.]

Sir, *Bristol, Nov.* 1, 1787.

I Take the liberty of communicating to the Society, such of my experiments in agriculture as I thought might merit notice, made within these ten years past—the term assigned by the Society in their Premium-book 1786.

Nothing

Nothing ſhort of an inſtitution, like that of the Bath Society, would be in any degree adequate to the due regiſtry of experiments in agriculture, and to their tranſmiſſion for the advantage of poſterity. Since the eſtabliſhment of that Society, the management of my farm hath been greatly directed to promote its general deſign; and yet I am almoſt aſhamed to conſider how barren I find myſelf of uſeful experiments.

The endeavours of an individual are very limited, frequently interrupted, and ſometimes wholly defeated, from a variety of occurrences; beſides that an experiment, conducted perhaps ſucceſsfully for months, or even for years, is probably loſt, in the loſs of only a few hours.

The cultivation of the land, as it is an employment the moſt innocent in its nature, ſo is it the moſt neceſſary and uſeful. It is the foundation and ſupport of all others. Trade could not ſubſiſt without it, and perhaps it is the only employment of which it may be ſaid, that the whole community flouriſhes in proportion to the proſperity of the individual engaged in it.

No. I. *Experiment on deep Ploughing.*

From the experiments and reaſoning of Monsr Chateauvieux, Du Hamel, and others, I was determined

determined to try the effects of deep ploughing; for this purpose I provided myself with a very stout plough, and began with a piece of land about 5 acres on Brislington Common, to which my experiments have been mostly confined. The soil a rich loamy sand, the colour that of a hazel-nut when fully ripe: the upper stratum of a pretty uniform quality, to the depth of from 2½ to 4 feet. In the first place, I had to contend with the prejudice of the ploughman, who, for what reason he knew not, very strongly objected to deep ploughing; however, I soon brought him to submission, and not without much labour he performed the business to my entire satisfaction:—the general depth was about nine inches.

During the operation, the ground was visited by most of the farmers in the neighbourhood, and the method universally exploded. By some it was said I should not plough for them, though I would do it for nothing; by others, that the land would not recover for seven years; and again, that it would be quite ruined. From all this I was not discouraged, and after giving the land two other ploughings, which were performed with ease and pleasure to the ploughman, it was cropped with Lammas wheat, and the produce was estimated, by some of these very farmers, at 40 bushels per acre. The next year it was manured with about 20 putt-loads of a composition, consisting of half hog dung, and half virgin earth, part of the same field, and planted with potatoes. This crop was kept free from weeds, well hoed and earthed up, the produce not less than 140 sacks, or 560 Winchester bushels per acre. Next it was sown with flax-seed, and produced two packs per acre, the pack 240lb. After this two succeeding crops of hoar-wheat, both good, say about 30 bushels each crop per acre.

The last season it was sown with black oats, and layed down with clover and ray-grass seed. The oats I estimate at 64 bushels per acre, which will appear moderate in comparison of an experiment upon that grain, which I shall note in the sequel.

I attribute greatly to deep and frequent ploughing, the success of these crops, and I persist in the same method. It is probable, however, that some degree of caution may be necessary on particular soils.

No. II. *Experiment on Turnips*—1782.

Four acres of ground were divided into two equal parts; one half manured with four putt-loads of soapers waste ashes, and the other remained without any manure. Turnip seed was sown on both

at

at the ſame time. The manured part proved an excellent crop, the other quite deſtroyed by the fly.

No. III. *On the Culture of Flax.*

The ſoil a rich loamy ſand, five acres, as per experiment No. I.

EXPENCES.	£.	s.	d.
Rent of 5 acres, at 40s. - - -	10	0	0
Two ploughings, at 5s. each - -	2	10	0
Sowing and harrowing, at 1s. - -	0	5	0
Fifteen buſhels ſeed, at 7s. - - -	5	5	0
Pulling the flax, at 8s. per acre - -	2	0	0
Watering and preparing, about 10s. per ditto	2	10	0
Swingling or dreſſing, &c. 203 dozen, at 1s. 8d. per dozen - - - - -	16	18	4
Ripling the ſeed, at 8s. per acre - -	2	0	0
Contingencies, at 5s. per acre - -	1	5	0
	£.42	13	4

N.B. It is to be obſerved, that I have not charged any of theſe experiments with tithe; the common being as yet exempt from tithe.

PRODUCE.	£.	s.	d.
Ten packs of flax, at 5l. 5s. - - -	52	10	0
Thirty-five buſhels of ſeed, at 5s. - -	8	15	0
	61	5	0
Deduct expences -	42	13	4
Profit -	18	11	8

or 3l. 14s. 4d. per acre.

The management of flax is tedious and difficult in these parts, by reason of the scarcity of proper workmen. Those we have are emigrants from the West, and take every opportunity of imposing on the inexperience of young farmers. Besides it appears to me that flax-growers ought to make it their staple article, and to consider the other parts of their farm as in subserviency to it. For the present, therefore, I have discontinued this culture. If I mistake not, there is a parliamentary bounty of 13s. 4d. per pack to the growers.

No. IV. *On Jerusalem Artichokes.*

At a considerable expence and trouble, I procured of these roots sufficient to plant half an acre of ground, but I have them now in great plenty. I find the produce to be about 480 Winchester bushels per acre; and I think they are about equal in value to potatoes for feeding store-pigs, such as are not less than five or six months old. For fatting hogs, I do not find they are near so valuable as potatoes. But their chief recommendations are, the certainty of the crop, that they flourish in almost any soil, and do not require any manure, at least for such a produce as I have stated. They are proof against the severest frost, and may be taken out of the ground as occasion may serve. Whereas potatoes are soon affected with frost, and must therefore

fore be ſecured before the winter ſeaſon ſets in. I generally plant three or four acres in a ſeaſon.

Expences per acre—drills 3½ feet aſunder, ſets nine inches—time, beginning of March.

	£.	s.	d.
Rent - - - - - - - -	1	10	0
Two ploughings, at 5s. - - -	0	10	0
Four ſacks of ſets, at 2s. - - -	0	8	0
Flat hoeing - - - - - -	0	2	6
Earthing up twice, at 2s. 6d. - -	0	5	0
Digging 120 ſacks, at 3d. - - -	1	10	0
	£.4	5	6

PRODUCE.

One hundred and twenty ſacks, at 2s. -	12	0	0
Deduct expences -	4	5	6
Profit -	£.7	14	6

No. V. *On Woad.*

Having been in converſation with ſome growers of woad, who reſide at Keynſham, a place famous for the manufacture of this valuable dye-ſtuff, it was aſſerted by them, that the growth of woad was peculiar to their ſoil and ſituation, and that the ſoil of Briſlington would by no means ſuit it; and indeed there is a very conſiderable difference in theſe; that of Keynſham, where the woad is raiſed, being a blackiſh heavy mould, with a good proportion of clay, but works freely; whereas the ſoil of Briſlington is moſtly ſuch as I have deſcribed it. I know of none ſimilar to that of Keynſham; yet I reſolved to prove how far their aſſertion was well founded; and obtaining ſome ſeed from one of the needy ſort, I ſowed half an acre on the common, an exceeding fine tilth, and a better crop I never ſaw in Keynſham. I could not prevail on any of the Keynſham growers to purchaſe it, although but about two miles diſtant; and not having either apparatus or judgment to manufacture, I ſuffered it to run to ſeed, gaining only from the experiment, that it is of a very eaſy culture, and might be made general, and that the only difficulty is in the preparing it for market.

No. VI. *On Coriander Seed.*

March 22d, 1783, ſowed ten perch with coriander-ſeed, the ſoil a good ſandy loam,

EXPENCES.	£.	s.	d.
Three ploughings - - - -	0	1	6
Sowing and harrowing - - -	0	0	1
Four pounds of ſeed, at 3d. - -	0	1	0
Harveſting - - - - -	0	0	3
Ripling - - - - -	0	1	0
Rent - - - - - -	0	2	0
	£.0	5	10

PRODUCE.

PRODUCE.	£.	s.	d.
87 pounds of coriander feed, at 3d. -	1	1	9
Deduct expences -	0	5	10
Profit	£.0	15	11

or 15l. 18s. 4d. per acre.

I have fince made feveral larger experiments in this article, but none has proved fo good a crop as the preceding; yet all of them fuch as to afford a good profit. There is a ready fale for it with the diftillers, druggifts, and confectioners. The former purchafe very large quantities—the price varies from 16s. to 42s.

No. VII. *On Brining Seed-Wheat.*

At my outfet in farming, I had frequently fmutty wheat, until, about feven years ago, I adopted the brining method, which, excepting in one inftance, I have invariably purfued, and faving that inftance my crops have been invariably free from fmut. The method is this:—

Mix falt with common water till it is capable of bearing an egg floating on its furface; introduce the feed wheat, well ftirring it about, fo that the light imperfect grains and other refufe matter may fwim at the top; this muft be carefully fkimm'd off

off from time to time till none ariſe; let it remain the ſpace of 12 to 16 hours; after which drawing off the brine at a ſpigot or cock below, placed there for that purpoſe, take away the ſeed grain, and after ſuffering the remaining moiſture to drain off a little, ſprinkle it with fine powdered lime, or wood-aſhes, it will then immediately be in a proper condition for ſowing. Although I remember that ſome years paſt, a ſudden and ſevere froſt interrupting our ſowing, ſome ſeed thus prepared remained ſo a full month, was afterwards ſown, and vegetated as perfectly as if it had been ſown immediately. The ſame brine will anſwer equally for any operations, and even for years, only ſupplying the loſs abſorbed by the grain.

No. VIII. *On Spring Wheat.*

April 9th, 1784, ſowed 1½ acre of ſpring wheat, the produce was 10¼ ſacks or 45¼ Wincheſter buſhels. It being an unuſual ſeaſon for ſowing wheat, brining was forgotten, and the crop proved remarkably ſmutty.

As to the practice of ſowing wheat in the ſpring, I am no advocate for it, only in caſes wherein the land cannot be got in order at the proper ſeaſon.

No. IX. A

No. IX. *A Comparison between Brining and not Brining.*

Of the ſmutty wheat in the laſt experiment, I cauſed a buſhel to be ſown unbrined, on half an acre of ground, and a buſhel brined on another half acre; the crop of the brined was free from ſmut, the unbrined very ſmutty.

No. X. *On recovering Smutty Wheat.*

I took a ſample of the ſmutty wheat [Experiment No. VIII.] to my baker, which he was very unwilling to purchaſe at any rate; at length, however, he offered me 16s. per ſack, 36 gallons: this was ſo much under the current price of ſound wheat, that I could not think of accepting his offer.

Some days afterwards it came into my mind to waſh and dry it; accordingly I provided myſelf with a tub conveniently ſhallow, that would well cleanſe about two buſhels at each operation, reſerving a ſuitable ſpace above the grain for the water, placing this under a pump; whilſt one man was pumping, another kept continually ſtirring it about with a broom, the ſmutty water, together with the light grain, overflowing the ſides of the veſſel, till the bulk of grain was thoroughly clean and bright. Thus in a few hours we compleated that part of the buſineſs. Next it was committed

to

to a malt-kiln for drying; and as I thought a much greater degree of heat ought not to be communicated to it than that of a hot ſummer, I never ſuffered it to exceed the 85th degree on Farenheit's thermometer, which I was well enabled to regulate by the application of that inſtrument. In the ſpace of about 18 hours the drying was finiſhed, and the whole performed greatly to my ſatisfaction;—not the leaſt veſtige of ſmut in ſmell or appearance. I then took a ſample from the kiln to the ſame baker, acquainting him with all the circumſtances; he was ſurpriſed at the metamorphoſis, and, after examining the bulk on the kiln, purchaſed it at 27s. per bag, confeſſing it was nothing inferior to any wheat of equal weight, the top of the market being then 28s. per bag; the loſs in meaſure attending the experiment was ſomething leſs than half a buſhel.

STATE *of the* EXPERIMENT, *viz.*

				£.	s.	d.
Ten bags of wheat, as per baker's firſt offer, at 16s. - - - - - - -				8	0	0
Ten bags ſold after the wheat was cleaned, at 27s. - - - - - - -				13	10	0
Deduct 2 men's wages, 2 days	£.0	6	0			
Loſs in meaſure half a buſhel -	0	1	8¼			
Fuel and rent, ſuppoſe - -	0	5	0	—0	12	8¼
				12	17	3¾
				8	0	0
Saved by the experiment -				£.4	17	2¾

No. XI. *On Canary Seed.*

March 1783, ſowed one peck of canary ſeed on half an acre of land, the ſoil a mixture of loam and clay—produce 8½ buſhels.

EXPENCES.	£.	s.	d.
Three ploughings, at 2s. 6d. - -	0	7	6
Sowing and harrowing - - -	0	1	6
Weeding - - - - -	0	4	0
A peck of ſeed - - - -	0	2	0
Cutting and harveſting - -	0	2	0
Threſhing, 9d. per buſhel - -	0	6	4½
Rent - - - - - - -	1	0	0
	£.2	3	10½
PRODUCE.			
8½ buſhels canary ſeed, at 10s. - -	£.4	5	0
Deduct expences -	2	3	10½
Profit -	£.2	1	1½

or 4l. 2s. 3d. per acre.

I have made ſeveral other experiments in this culture, but never exceeded the above in produce; although it is ſaid that in the Iſle of Thanet, where this crop is not unfrequent, they uſually obtain upwards of 20 buſhels per acre.

No. XII. *On Aniſe.*

I have tried ſeveral experiments in the culture of aniſe, but was never fortunate enough to get a

crop,

crop, it appearing that this climate is not in general ſufficiently warm to mature and perfect the ſeed; the diſtilled plant however, uſing it when in bloſſom, affords a more ſweet and grateful tincture than either the ripe ſeed or eſſential oil.

No. XIII. *On Potatoes.*

The quantity of land 6½ acres, a mellow, deep, ſandy loam, on Briſlington Common—diſtance, drills three feet aſunder, ſets eight inches.

	£.	s.	d.
Three ploughings, at 5s. - -	4	17	6
Thirty-five ſacks of ſeed potatoes, at 5s.	8	15	0
Planting, at 3s. 6d. per acre - -	1	2	9
104 putt loads of manure; compoſition, 2-thirds natural mould, and 1-third hog-dung, at 2s. - - - - - -	10	8	0
Taking out of the ground - -	13	0	0
Bringing to market - - -	13	0	0
Three hoeings, at 12s. per acre -	3	18	0
Rent - - - - - -	13	0	0
	£.68	1	3

PRODUCE.

Potatoes ſold - - - -	£138	5	0
Ten ſacks uſed in the family, at 5s. -	2	10	0
Sixty ſacks reſerved for planting, at 5s.	15	0	0
	155	15	0
Deduct expences -	68	1	3
Profit -	£.87	13	9

or 13l. 10s. per acre nearly.

Besides, the whole expence of manuring ought not to be charged to this experiment, the succeeding crops clearly evincing the great advantage they received from it; for it is worthy remark, that this piece of land never received but the single dressing mentioned above to this time, and yet has produced stout crops of wheat, and potatoes alternately, until last spring it was laid down with grass seeds, and sown with oats.

No. XIV. *On Black Oats*—1787.

The same land as in the last experiment. The preceding year it had carried potatoes, and received one ploughing for a winter fallow.

In February last, another ploughing was given, and on the 27th and 28th of the same month four Winchester bushels per acre of black oats were sown; this was earlier by about a month than oats are generally sown in our parish, and I did it with a view to ascertain the effects of early sowing. When the oats were ripe, I caused exactly half an acre to be cut with the sickle, and sheaved; these were threshed out, the produce was 49¼ Winchester bushels—a quantity most amazing in these parts. The success of the crop I impute partly to early sowing, and partly to good deep tillage; and I believe the half acre was a fair average of the whole piece.

EXPENCES *of an* ACRE.	£.	s.	d.
Rent - - - - - - -	2	0	0
Two ploughings, at 5s. - - - -	0	10	0
Cutting - - - - - -	0	3	0
Harvesting - - - - - -	0	5	0
Four bushels of seed, at 2s. 6d. -	0	10	0
Sowing and harrowing - - - -	0	3	0
	£.3	11	0

PRODUCE.

	£.	s.	d.
98½ bushels of oats, at 2s. - - -	£.9	16	6
Deduct expences -	3	11	0
Profit -	£.6	5	6

The straw may be valued in lieu of threshing, conveying to market, &c. but is worth abundantly more than what would defray those expences.

I am, Sir, your obedient servant,

NEHEMIAH BARTLEY.

ARTICLE XXXI.

On the Black-Rust in Wheat.

[In a Letter to the Secretary.]

SIR, *Bradley-House, July* 27, 1785.

THE bearer hereof, RICH. WINSOR, of *Berry-Pomeroy*, near Totnes in the county of Devon, yeoman, has found out a method of curing the Black-

Black-Ruſt in wheat, which he has tried ſeveral ſeaſons, and found it to be of great utility; and others, who have taken his advice in trying the experiment, have likewiſe reaped a conſiderable advantage by it.

The method he has found out for curing it, is to let ſuch ruſty wheat ſtand uncut, three weeks or more after the uſual time at which people in general cut ſuch wheat.

He attributes the infection to ſmall inſects, falling upon the ſtalk in foggy or miſty weather; inſects of a poiſonous nature, that cauſe the ſtalk to ſwell, and the knots of the ſtalk to cloſe; by which means the ſap, which ſhould go to nouriſh the grain, is prevented; and that by letting it ſtand as aforeſaid, the ſun and air will deſtroy theſe inſects; the knots will then open, and as they open, the ſap paſſes up and feeds the grain; which, by letting it ſtand a proper time, will recover and become much more full, and will be near as good in quality, as though no ruſt had happened to it.

Mr. Winsor acquainted the Society in London of this matter ſometime ago, not knowing of a Society at Bath, till I informed him of it; and I have recommended him to apply to you as their Secretary,

Secretary, desiring you to lay it before the Society, who, he doubts not, will reward merit according to its desert. I am, Sir, with all due respect,

your humble servant,

RICHARD BAKER.

*** The foregoing short account, stated as a matter of fact founded on experiment, we give to our readers for their consideration; and as further experiments cannot be attended with any probable disadvantage, we conclude the method will have a fair trial among those farmers and gentlemen to whom the hint may be new.

Article XXXI.

Recipe for making Rennet for Cheese.

[In a Letter to the Secretary.]

Sir, *Frome, Oct. 5, 1787.*

AMONG the various subjects which engage the attention of the members and correspondents of the Bath Agriculture Society, it appears somewhat strange, that the two grand articles within the province of the Dairy-woman (Cheese and Butter) have not been more attended to.

The Agriculturist has been repeatedly informed of the proper management, the best manures, and the

the likelieſt crops, which may be applied to each reſpective ſoil: but the good houſewife, the ſedulous dairy-woman, who daily furniſhes us with two of the chief ſupports and luxuries of life, has been left to grope out her way, through this age of improvement, with the little ſtock of knowledge which, in early life, ſhe imbibed from her mother. I wiſh, therefore, that the members of your Society would now and then beſtow a little of their attention on theſe good women, who ſo much want and ſo highly deſerve it.

It is not within the compaſs of a letter, that inſtructions can be fully given for making cheeſe and butter; yet, as detached obſervations on thoſe ſubjects may ſometimes have their uſe, I ſend you a recipe for making rennet for curdling cheeſe.

Take the *abomaſa*, commonly called the vells or pokes of calves, killed before they have fed on vegetables, and waſh them in clean water, ſalt them well, and lay them in ſalt for two months; then, with the ſalt about them, hang them up in a coarſe bag in the chimney (not too near the fire) for ten months. In the ſpring following, when the cowſlip is in full bloom, gather a quantity thereof, and pick the petals from the calixes, and boil them in a ſufficient quantity of water for a quarter of an hour,

 with

with the proportion of a pound of ſalt, and an ounce of allum to every twelve pints of water. Let this brine ſtand to cool until the next day, when it may be ſtrained off from the cowſlips. To every gallon of this brine, put in two pokes, and let them remain four days, at which time you may bottle it off, putting two or three cloves and as many grains of allſpice into each bottle. Let the bottles be corked tight, and the rennet will keep good a year or more. Two large ſpoonfuls of rennet, thus prepared, will coagulate a hogſhead of milk.

After the pokes have been thus uſed, let them drain dry, and ſalt them afreſh for a fortnight, and they will ſerve again, nearly as well as before.

Should this paper be found worthy of admiſſion, in the fourth volume of the Society's ſelect papers, I may be induced, at a future opportunity, to give you ſome further thoughts on cheeſe-making.

I am, with reſpect, yours, &c.

A. CROCKER.

Article XXXIII.

On the Benefit of Cultivating Parſnips and Burnet.

Gentlemen,

I Have with much pleaſure and much inſtruction peruſed your ſelection of papers communicated to the public; and am of opinion, that there is a plant, I mean the Parſnip, which has not been yet tried by any of your correſpondents; but which is in France, and in our adjoining iſlands, held in high eſteem as a food, particularly for cattle and ſwine. In Brittany, eſpecially, they mention it as little inferior in value to wheat. Milch cows fed with it in winter, ſay they, give as good milk, which yields as well-flavoured butter, as milk in May or June, and in as great abundance. It is much commended for ſwine, which rear young pigs. It alſo proves very uſeful in fattening ſwine.

For a complete account of its uſes, conſult a volume of Memoirs publiſhed by a Society at Rennes, inſtituted for ſimilar purpoſes as your's. I think there is a tranſlation of the Memoirs in Mills's huſbandry.

Some judgment may be formed of the comparative value of plants as food, from the proportion of mucilage they contain, or yield in decoction; for this purpose, suppose a pound weight, for instance, of parsnips, carrots, potatoes, &c. were boiled separately in a quart of water, the decoction strained, and, when cold, compared. The decoctions of the parsnips will, I believe, be found the most mucilaginous, or the most thickened. Be this as it may, the culture and trial of the plant seems an object worthy the attention of your Society.

Farmers are apt to judge of the merits of plants by the weight of their productions, without attending properly to their different qualities. Thus *Burnet* is, I find, by your correspondents made little account of. Upon trial it will be found that it goes much farther in feeding sheep, for instance, than any other plant. Thus, suppose that some sheep are fed on an acre of it, and an equal number on an acre of any other plant; I have some authority to say, that sheep will be longer well fed on burnet, than on any other plant I know. The mutton of sheep fed on it will be better coloured, more juicy, and better flavoured, than the mutton fed on any other food. It stands the winter better, and shoots as early in spring as any plant. It has been found to be a perfect cure of the *rot* in sheep; and cows, sheep, or goats, fed on it, give more milk, and more nourishing milk, than on any other pasture;

pasture; and the butter obtained from their milk is not inferior to any.

I have mentioned sheep particularly, because burnet seems to be more peculiarly beneficial to them than to cattle.

The great excellence of the Turnip-rooted Cabbage is, its being a certain and early food in spring, when it is generally most wanted.

Wishing your Society the success they so well deserve,

I am, with much respect, Sir,

Your obedient servant.

A Lover of Georgical Pursuits.

N. B. We agree perfectly with our correspondent, in a high opinion of the value of parsnips, as a food for cattle; and have been induced to insert his letter as a fresh call of the public attention to the subject, though by no means as to a new, or wholly neglected matter. His encomium on burnet may be considered also as much anticipated by former writers. But if it shall be proved, that this well-known plant is either generally, or under particular circumstances, a perfect cure for the *rot in sheep*, much benefit will be found to result from the fact.

Article XXXIV.

On the Use and Value of Turnip-rooted Cabbage.

[In a Letter to the Secretary.]

Sir, *Hethel, June* 21, 1787.

I Have been for a long time so much occupied by other matters, that little leisure has been afforded me for experiment or observation on agricultural affairs. The following one, whereby the use and value of the Turnip-rooted Cabbage may be in some degree ascertained, I transmit for the inspection of the Gentlemen of the Society, and submit to their consideration, how far, from this account, the cultivation of that root appears to merit their future encouragement.

The following is an account of the cattle or beasts fed from five acres of turnip-rooted cabbages; four acres of which were eaten upon the land as they were growing, (but parted off by fold-hurdles into portions of about an acre each) and one acre pulled up and carried to the stables and ox-houses. These turnips were sown and cultivated as other turnips; the beasts were put to them on the 13th of April, and continued feeding upon them till the 12th of May following.

Twelve

	£.	s.	d.
Twelve Scotch bullocks, weight 40st. each 4 weeks, at 2s. per head per week - -	4	16	0
Eight homebreds, 2 years old, at 1s. ditto -	1	12	0
Fifteen cows full-sized, at 2s. per week -	6	0	0
Forty sheep, at 3d. ditto - - - -	2	0	0
Eighteen horses, fed in the stables with an allowance of hay, at 1s. ditto - - -	3	12	0
	£.18	0	0

Besides 40 store hogs and pigs, which lived upon the broken pieces and offal, without any other allowance for the whole 4 weeks.

When it is considered, how very nourishing a food the turnip-rooted cabbage is, the price I have fixed to the keeping each beast per week will not, I conceive, be deemed too high. I am sure the farmers here will always, at that particular season of the year, be willing to give it, and more; because it enables them to spare the young shooting grass (which is so frequently and greatly injured by the tread of the cattle in the frosty nights) until it gets to such a length and thickness as to be afterwards but little affected by the drought of the summer. They have besides other great advantages to recommend them to a more common use; they are never affected by the most intense frosts; if bitten by sheep, hares, rabbits, or the wood-pigeons, (which in this place abound to the great destruc-

tion

tion of turnips near any woods) they hardly ever rot. The tops or leaves are in the ſpring much more abundant, and much better food than thoſe of the common turnip, and they continue in full perfection after all other turnips are rotten or worthleſs.

With theſe circumſtances to recommend them, it muſt however be owned, that they have inconveniences attending them. They require a great deal of time and pains to get them out of the ground, if pulled up to be carried elſewhere:—and if fed as they grow, they are ſo deeply rooted in the ground, that it requires the ſame labour to get the pieces out of the ground, and they riſe with abundance of earth entangled in the fangs of the roots. They are likewiſe ſo firm and ſolid, that the whole ones, when pulled up, require to be cut in halves, that the cattle may be enabled to eat them.

To obviate ſome of theſe objections, it will be proper to ſow them on rich and very light land; and as they are longer after being ſown in coming to the hoe, than the common turnips, I have found it neceſſary to ſow them earlier, ſo early as the beginning of June.

I have grown them a great number of years; from the experience I have had of their utility I continue to cultivate them; and I think no gentleman, who keeps them to consume for the last fortnight or three weeks before he turns his cattle to grass, will have reason to grudge the expence or trouble attending them.

If in any enquiry or other business here I can be of the least use, you may freely command me, and I shall be proud on every occasion to shew that

I am, Sir,

your most obedient servant,

THOMAS BEEVOR.

ARTICLE XXXV.

On the Mangel-Wurzel, or Scarcity Root.

[By the SAME.]

SIR, *Hethel, Oct.* 12, 1787.

I Feel myself highly flattered by the favourable opinion the Gentlemen of the Society are disposed to entertain of those accounts in husbandry, which it has been in my power to send them; and give me leave to say, that I have particular reason to be pleased with the polite and friendly manner in which you have expressed their and your approbation of them.

I wish

I wiſh I could by any freſh communication convince them, that I was deſerving of their commendation; but from the many and various avocations I have lately had, I have been rendered leſs able than I could wiſh, to attend to any experiments worth relating.

I this ſummer received from a friend who came from Paris, ſome ſeeds of the plant called in Germany *Mangel-Wurzel*; by M. DE COMMERELL, *Racine de Diſette*; and in Engliſh, *the Scarcity-Root*. The account of the plant, and its time and method of propagation, are ſo fully given by M. L'ABBE DE COMMERELL above-mentioned, in the Memoire publiſhed by him, and which I ſuppoſe you have ſeen, that I ſhall wholly omit the mention of them, and only relate what little I have yet obſerved of it.

My ſeeds were ſent me very late, two months nearly after the moſt proper time of ſowing them; however I ventured to commit them to the ground on the 12th day of June laſt; and in a few days had the ſatisfaction to find them all riſe well, and in a vigorous ſtate of growth. I have ſince gathered their leaves twice, and find their roots of ſuch ſize as to promiſe a conſiderable and profitable production. The meaſure of ſome of them is now 15 inches round; the length (of a few I pulled up on this

this occasion) is 13 inches, and the weight of them on an average 4lbs. The seed and plants are not, I think, to be distinguished, at their first growth, from some beets; but in order to ascertain the difference, (if such there was) I sowed on the same bed of mould, on the same day and hour, some seeds of the real beets; and find that, under the same management, the roots of the scarcity plant are four times as big, and the leaves of it much larger than those of the real beets. I have offered a few of the leaves of the scarcity plant to the cows whilst going in exceeding good pasture in my park, which they readily ate; I did the same to some horses which were standing in a waggon in the harvest field, who as readily ate the broad tender part of the leaves, but rejected the thick parts of the stalks. I have also had dressed the leaves of each of the above-mentioned plants, and brought boiled to my table; and think, as did some other gentlemen who ate of them, that there is a manifect difference in their taste; those of the scarcity plant being so like spinage, as hardly to be distinguished from it; whilst those of the beet were both harder and drier.

What further observations I shall be able to make upon the growth and application of this plant, in the course of the winter, I will transmit to you,

you, as it certainly promiſes to be of the firſt importance in the article of food for cattle. In the mean while let me not omit to inform you, that I ſaw a few weeks ago at Lord ORFORD's place at *Eriſwell*, near Barton-Mills in Suffolk, ſome of the plants, which were nearly twice as big as mine:—and I have been told, that at Mr. DASHWOOD's, of *Cley*, in Norfolk, there are ſome which meaſure two feet in circumference; but the two laſt-mentioned parcels were, I am informed, ſown at leaſt ſix weeks ſooner than mine were.

I have had ſent me this week an account of a moſt wonderful production of vetches: upon two plants ſown in the garden of JOHN BERNEY PETRE, eſq; of *Weſtwick*, in Norfolk, there were found (after ſeveral had been accidentally plucked off) no leſs than 994 pods, containing on an average ſix ſeeds in each pod; in all 5964 ſeeds.

Mr. PETRE, who ſent me the account, did not know the name of the plants; but from a branch of it which he ſent me, with the account, I have great reaſon to believe it to be the broad-leaved many-flowered vetch of Crete; for it had upon it ſome deep purple flowers, and is a perennial plant, as he aſſured me; however, not having any bota-

nical

nical book by me at present, I cannot be at all certain of the truth of my conjecture.

P. S. In riding yesterday about seven miles from this place, I saw at least two-thirds of the wheat for next year's crop was dibbled, and set by hand. I am inclined to believe it will soon be generally so here.

Article XXXVI.

Experiments on various Sorts of Potatoes.

[By the same.]

Sir, *Hethel, Dec. 1, 1787.*

I Venture to send you an account of a trial made by me, of a few sorts of Potatoes planted last spring; and as there is not, perhaps, in the wide field of agriculture, any plant which more deserves attention and general cultivation than the potatoe; so I hope every information which leads to the discovery of the best and most productive kinds, will be received in good part, and neither deemed trifling nor useless by any of those who are real well-wishers to the interest of society.

I shall content myself with this apology for the contents of my letter; and after premising that all the underwritten potatoes were planted on the 2d day of April, in a garden, the soil of which is a rich hazelly

hazelly coloured loam, neither too wet nor too dry; that they were all well dunged, for that the four firſt ſorts ſtood on a border where a row of apple-trees had grown, which were taken up about a month only before the ſets were planted; that the three laſt ſorts were planted on ground which had been cropped as gardens uſually are, and that the pieces planted were cut from large potatoes, with two or three eyes on each piece; I will proceed to ſhew the reſult of the experiment.

No.	Names.	Weight of ſeed.		Quantity of Ground.	Weight of produce.		Buſh. per acre.
		lb.	oz.		ſt.	oz.	
1.	Incomparable, a ſeedling,	4	9	6-10ths of a rod	13	0	692
2.	Denne's Hill, dit.	3	1	8-10ths	16	10	668
3.	Bayley's ſeedling,	3	1	5-10ths	8	6	539
4.	Manley White	4	12	3-10ths	6	4	670
5.	Kentiſh ſeedling,	2	10	4-10ths	16	11	1342
6.	Champion,	3	6	5-10ths	11	1	708
7.	Ox-Noble,	3	11	4-10ths	14	0	1140

The above roots were all taken up on the 29th day of October laſt, and the ſtems of each, except thoſe of the Manley and Champion, which were entirely dead, were green and freſh at the time of taking them up.

No. 1, large white, meally, ill-taſted.

No. 2, very large, white, meally, and good.

No. 3, middle-ſized, white, meally, and exceeding well flavoured.

No. 4, large, white, meally, and good-taſted.

No. 5, very large, white, not yet tried upon the table.

No. 6,

No. 6, middle-ſized, white, meally, and exceeding good to eat.

No. 7, large, white, and ill-flavoured.

The buſhels above-mentioned are heaped buſhels, weighing on an average 70lb. per buſhel.

Article XXXVII.

On Planting of Waſte Lands.

GENTLEMEN, *Norwich, Feb.* 29, 1788.

THOUGH planting waſte land be not immediately within the province of agriculture, yet the publick advantage, of which you are the promoters, may be more effectually ſerved by the ſtudy of certain modes of planting it, than from annual crops; and eſpecially as planting becomes eventually a uſeful auxiliary to cultivation. I therefore wiſh to preſent to your notice, as a poſſible example to other parts of the nation, the practice and ſucceſs of a neighbouring gentleman (Sir Wm. Jerningham) on the moſt unpromiſing ground, perhaps, that any ſucceſsful planter has hitherto attempted; notwithſtanding there is a certainty from experience to believe, that the ſtubborn ſoil may be meliorated, and the apparent ſterile be made productive; and by properly timing the period for ſpecifick

ſpecifick productions, what would in the natural ſtate of the land have been impoſſible, by an adaption of fit circumſtances, a production may be excited, foreign and uncongenial to the ſoil; while, without theſe circumſtances, no more ſucceſs would have followed than to him who ſows without culture, or plants without trenching the ground.

Theſe reflections aroſe from a frequent and late obſervation, made on the extenſive and thriving plantations of the abovementioned gentleman; who has, without hyperbole, changed the barren heath to a fruitful field, the dreary waſte to a delightful foreſt, by an adaption of circumſtances to ſituation and ſoil; and, by an application of what would cheriſh and defend, has extended a plantation of beech-trees uncommon to this diſtrict; nor I believe do they ſpontaneouſly grow in any county through the eaſtern diviſion of the kingdom.

The mode Sir William purſued, was the planting of the beech-trees from the nurſery, while ſmall, amongſt Scotch firs. Many heaths beſide his have been broken up and planted with firs, to much publick and private benefit. But I have not obſerved, unleſs recently, the regular intermixture of the beech at due diſtance. Theſe trees, in a ſoil perhaps without clay or loam, with the heathy ſod, trenched into its broken ſtrata of ſand or gravel, under the protection of the firs, have laid hold, though ſlowly, of the ſoil, and, accelerated by the ſuperior growth of the firs, have proportionally riſen, until they wanted an enlargement of ſpace for growth, when the firs were cut down.

It is ſcarcely neceſſary to obſerve, that when this reſinous tree is felled, the roots decay in the ground, and furniſh by that decay a new ſupport to the ſoil on which the beeches grow; by which contingency, they receive an added vigour, as well as the favourable concomitants of an enlarged ſpace in earth and air; and by being now diſincumbered from their former ſupporters, their growth becomes more and more obvious; they are ornamental to the country, promiſe in time to be uſeful timber, and probably may diſſeminate their ſpecies where they would not have been expected to flouriſh, but under the ſhade and encouragement of the firs. And theſe firs having met with no obſtacle from the infant timbers they encouraged, their boles are now converted to poſts, rails, and various other uſes, and their branches have been bound into thouſands of bavins, that have heated the ovens, or have been burnt on the hearths of the farmers and cottagers around.

JOHN WAGSTAFFE.

Article XXXVIII.

Description of a Model of a Machine for communicating Motion at a Distance.

[In a Letter to the Secretary.]

SIR, *Bristol, Nov.* 21, 1786.

A Commodious method of communicating motion at a diſtance, has long ſtood among the deſiderata of mechanics, and no method that I know of has been attempted, that at the ſame time will free the machine from weight and incumbrance. Hollow ſhafts of caſt-iron bid faireſt to anſwer this end, but they are expenſive; and though I ſuppoſe it may be cheapeſt, all things conſidered, it will be difficult to make them general.

The method introduced to you by the model accompanying this letter, was tried and approved of at a mill in the neighbourhood of this city, on two ſhafts, one 15 feet long and 15 inches thick, and the other 12 feet long and 12 inches thick; the latter of which was ſo weak as to twiſt near a tenth of its circumference; which, when the reſiſtance became in any wiſe unequal, ſubjected the whole machine to the greateſt danger and diſorder.

It was propoſed and intended to take it out and ſubſtitute a larger, which would have been attended with

with great expence and inconvenience, as the whole manufacture muſt have been interrupted for ſome time, and part of the mill-houſe muſt have been taken down. Being apprized of this intention, I adviſed the method here recommended. Accordingly two flat bars were procured from a ſcrap forge, of 2½ inches by full 3-8ths of an inch, and their ends furniſhed with ſcrew-pins of 1¼ inch, with ſquare threads: the bars were then hollowed on the under ſide with a large ſwage, in order to make the edges lie cloſe to the ſhaft. This done, they were annealed and put to a large vice, and twiſted with a hand-hook, ſuch as the anvil-ſmiths uſe; ſo that one end had made a little more than a revolution, after which a few blows of the hammer (and which requires not ſo much ſkill as may be imagined) formed them into a ſpiral, fitted to a cylinder of 12 inches diameter. They were then carried to the mill at the diſtance of five miles, and after the blocks were fitted to receive their ends, were put on with the greateſt facility.

It may be neceſſary to obſerve, that the diſtant gearing of the mill was trigged, while the water-wheel was turned back in order to twiſt the ſhaft the reverſe way of its going; by which means the ſpirals bound cloſer than could poſſibly be by ſcrewing only. I ſhould add, alſo, that ſome blows of a hammer, of about 12lb. were laid on in order to close them to the sides of the shaft; which being an octagon, and not a cylinder, could not be effected by any other means.

This was the method practised in strengthening those shafts already in gearing; but if it should be thought expedient to use them in construction, they may be applied to more advantage; for instead of one revolution of the spiral, it may have two; in which case the advantage will be double: add, that in both cases there may be as many spirals as there are arms in the wheel.

I am, Sir, your obedient servant,

J. C. HORNBLOWER.

N. B. It has not been thought adviseable to attempt any representation on a plate, of the model accompanying this letter; but our mechanical readers, who have any curiosity to see it, may be gratified, by applying at the Rooms of the Society.

Article XXXIX.

Sir, *West-Monckton, March* 6, 1788.

BEING struck very forcibly with the importance of the subject, I sometime ago committed the inclosed thoughts to writing, not at that time with any

any design that they should appear in publick; but I happened to shew them lately to a gentleman of the neighbourhood, who desired I would send them to your Society. I have, therefore, taken the liberty to follow his advice; and if you find any thing worthy the notice of the Society, or yourself, it will afford pleasure to, Sir,

Your obedient humble servant,

T. PAVIER.

"Were the forest of Dean duly improved, it were "an imperial design: and I do pronounce it "more worthy of a prince, who truly consults "his glory in the highest interest of his subjects, than that of gaining battles, or subduing a province: for he not only secures the "strength and glory of the nation, in preserving an abundant supply of timber for shipping; but also adds greatly to the number "of people, by the many new farms for corn "and grass, erected where the land turns at "present to little account for timber, which is "universally neglected; and less for men, being uninhabited."

The above is a quotation from Mr. EVELYN'S SYLVA, which a late survey of the land therein mentioned, and the ſmall quantity of timber reported to be now growing thereon, have brought afreſh to my memory. Such an improvement as is above recommended, would undoubtedly be of the greateſt importance to this kingdom, in future generations; and would redound highly to the honour and glory of a Britiſh government that ſhould carry it into execution, at the ſame time that it would be attended with but an inconſiderable expence.

To illuſtrate this aſſertion, let us ſuppoſe that inſtead of diſpoſing of all the waſte lands belonging to the crown, ſome particular places, where the ſoil and ſituation ſeem adapted for producing good oak timber, were reſerved to be improved for that purpoſe; the expence of incloſing is then the firſt thing that comes under conſideration; and this expence will always vary in proportion to the form and magnitude of the land to be incloſed; as a field of a hundred acres may ſometimes be fenced in for the trifling ſum of about 10s. per acre, whilſt another of but ten acres ſhall coſt by the acre three times as much.

It may happen that the ground for ſuch intended improvement may adjoin to lands already incloſed, which will greatly leſſen the expence; but in order to make ſome kind of calculation, I will ſuppoſe

ſuppoſe it to be fenced quite round on every ſide, and to coſt on an average 20s. per acre.

The next ſtep will be to prepare the ground for the reception of the acorns, which will undoubtedly be beſt effected by frequent ploughings; I would therefore propoſe to keep it in conſtant tillage for two or three years, till the earth is brought to a fine, mellow ſtate of tilth, and then to ſow or plant the acorns in the autumn: the profit of the crops taken from off the premiſes will (no doubt) abundantly overpay all the expence of incloſing, as well as the collecting and ſowing the acorns.

Theſe crops could not impoveriſh the ground ſo as to occaſion any injury to the intended plantation, becauſe the young trees will derive their nouriſhment and ſupport from that part of the ſoil which lies beneath the action of the plough, or the extenſion of the roots of any kind of corn.

From hence it appears, that an improvement of this nature would be attended with very little, or perhaps no expence, but what would be amply repaid by the profits ariſing from the ſame; and I am perſuaded, that no further trouble or expence would be neceſſary for ſeveral years, but to take care that no kind of cattle whatever be admitted into the incloſure.

When the plants are about eight or ten years old, it may be neceſſary to cut down the greater part, leaving a ſufficient number of the moſt promiſing ones, the ſuperfluous branches of which ſhould then be taken off, which ought to be the only time they ſhould ever be pruned; as it would be better for the young trees, that ſuch branches ſhould be ſtripped off by hand every time the underwood may be cut, as long as they can be eaſily bent down for that purpoſe, or if convenient, every year.

At every time of cutting the underwood, the young trees ſhould be thinned with great diſcretion; the thicker they ſtand in reaſon whilſt young, the better lengths will they arrive at: but I apprehend that each tree ſhould at the laſt be allowed a hundred ſquare yards for the expanſion of its limbs.

Allowing this to be a proper diſtance, an incloſure of fifty acres would produce 2420 trees, which I ſuppoſe would come to perfection in about 100 years, and that they would be worth (on an average) five pounds each; the value of the timber on the fifty acres would then be 12,100 pounds.

Mr. Evelyn computes the profit of a thouſand acres, in a hundred and fifty years, to amount to upwards of 670,000l. How he made ſuch a calculation,

lation, I cannot gueſs, but think the profit is charged much too high.

Oak timber, let up in the manner above deſcribed, would arrive to great lengths; and having never been pruned or tranſplanted, there could be no danger of their falling unſound; conſequently the charge I have made of 5l. a tree for their value on an average, will, I ſuppoſe, be thought very reaſonable, eſpecially as the bark and wood are both included.

The quotation from Mr. EVELYN, with reſpect to the foreſt of Dean, is moſt undoubtedly applicable to many others of the waſte lands belonging to the crown, and in particular to the New-Foreſt in Hampſhire, which would produce an immenſe quantity of fine timber without any expence, if a method could be deviſed to prevent the deer and other cattle from cropping the young trees in their infancy: the truth of this appears from an obſervation I made ſome years ago, that there was ſcarce a young oak to be ſeen but what had found its way up through a thick buſh of thorns or brambles, and conſequently owed its preſervation thereto: from this obſervation alſo, I am of opinion, that there can be no neceſſity for any conſiderable expence in weeding a young plantation of oak.

Complaints of the ſcarcity of oak-timber fit for ſhip-building, are at preſent very frequent; and from the ſmall quantities that are coming up in moſt parts of the kingdom, it ſeems to me very apparent, that ſuch ſcarcity will be ſeverely felt in another century; conſequently the preſent opportunity for making ſome ſuch improvements for the benefit of futurity, is highly deſerving the notice and conſideration of thoſe in power.

If the foregoing conſiderations are juſt and reaſonable, what vaſt advantages might future generations derive from judicious and provident improvements of this kind! And I am perſuaded, that ſuch undertakings would be recorded in hiſtory in terms that would oblige poſterity to look back with gratitude and applauſe to the period that produced them.

The foregoing reflections, however brief on a copious and national ſubject, are equally ſeaſonable, and fraught with importance. The improvement of any country in thoſe articles of produce, which are of greateſt conſequence to its ſafety and accommodation, is among the firſt objects of its provident care. In determining what thoſe articles of produce are, regard muſt be had to natural circumſtances of ſoil, climate, and ſituation, with reſpect to ſurrounding countries.

According

According to the ſtate of Europe, and the inſular ſituation of this country, much of its ſafety and importance have been politically determined to ariſe from a plentiful growth of oak timber, fit for building ſhips of defence, and for merchandiſe. Nor is the cultivation of a tree ſo congenial to our ſoil, and ſo ornamental to our foreſts and fields, an object unconnected with domeſtick uſes, in the conſtruction of various kinds of machinery, and the moſt firm and comfortable habitations.

We agree with Mr. PAVIER, that the cultivation of Oak-timber, with a view to the benefit of poſterity, ſeems to have been of late years too much neglected in theſe kingdoms. And while every friend of human felicity muſt condemn the miſguided ambition of a tyrannical prince, who could deſtroy whole villages to plant a foreſt, he will feel due ſolicitude for thoſe general advantages which muſt reſult from better maxims of cultivation.

It may not be deemed ſo fully within the province of this Society, to call on Government for its attention to the management of Royal Foreſts and Waſte Lands, as to point out the benefits of a general improvement on the eſtates of individuals. Such a ſpecies of improvement will at leaſt be liſtened to, as a worthy and proper object of rural œconomy; and every exertion that may be excited by ſuch means, will have ſome favourable influence in a nation emulous of greatneſs and of fame.

From this motive, we ſhall not heſitate to ſuggeſt the eaſy and manifold advantages that would reſult from increaſing plantations of oak-trees, on particular parts of numerous eſtates, from the lordly park, down to the ſmall cultivated farm. On the former, the growth of the oak

is

is truly deemed essential, both to the elegance and grandeur of the inclosure. But while this idea prevails in theory, and neither grandeur nor elegance can be realized without it, there is too much reason to fear that planting has not kept due pace with the consumption of this valuable species of timber. On the latter, it is much to be questioned, whether a view to *immediate profits* from the soil, has not too generally obtained to the exclusion of timber plantations.

Why this error should obtain, it may be difficult to determine in a way favourable to the wisdom and foresight of a multitude of land-owners. For on many inclosed farms, and especially on farms which have a bleak northern exposure, it would frequently be found that a judicious plantation of young timber trees would gradually increase the value of the lands, by the shelter they would afford to cattle, the strength they would give to fences in which they might be planted, and the breaking of unfriendly North and North-east winds.

But admitting the situation of inclosures to be such, as sometimes not to stand in any great need of the shelter of trees, it frequently happens, that on farms of considerable size, and variety of soil and exposure, small parts, of no great value for pasture or cultivation, might be appropriated, without any material lessening of the annual income of the farm, for entire plantations, even of oak. But if it should be thought that an entire oak plantation would be too great a sacrifice of the ground, a plantation of various other species of trees of a quicker growth, to be periodically cut as underwood, might be made, and the produce come in aid of the supposed disadvantage; while the principal object above contended for, would be secured.

To theſe conſiderations may be added, the ſource of *fencing*, hurdling, draining, and firing, (that would be created on many farms where thoſe conveniences are much limited, to the no ſmall inconvenience of the farmer) and the picturesque beauty, that ſuch plantations, generally adopted, would give to the face of a country.

Article XL.

On the Healthineſs of managing Silk-Worms.

[In a Letter to the Secretary.]

Sir, *Bridge-North, Cann-Hall, Dec.* 15, 1787.

THE life and changes of a Silk-Worm may juſtly be claſſed among the moſt wonderful phenomena of nature: and never have my ideas of the great Creator of all been raiſed to a higher pitch of enthuſiaſtic adoration, than whilſt contemplating this induſtrious little animal, excluded from light, from air, and ſuſtenance—and yet weaving, with mathematical exactneſs, the web which ſhall clothe the higheſt order of the world's inhabitants.

The incongruity of believing that Almighty Goodneſs could make that creature pernicious to man, for whoſe ſervice and delight he is evidently created, will be ſufficiently obvious to you;—but vulgar prejudices muſt be combated with other proofs.

If

If the fact were really so, what would become of the inhabitants of Italy, of China, and more especially of the islands in the Archipelago; where, from the immense numbers which are reared, the whole atmosphere must be impregnated with their deleterious effects? I have been assured, by an intelligent friend, who spent some time in Italy, that whenever they had epidemick complaints, the children who had the care of the silk-manufactories invariably escaped the contagion; and this I have reconciled on the principles of Dr. PRIESTLEY, who asserts, that the air in rooms is rendered doubly salubrious, by the introduction of opening vegetables, or fresh-gathered leaves.

The silk-worm in itself is totally inoffensive; but if dead ones are suffered to remain among them, they certainly become putrescent, as other animal substances, and of course unwholsome.

In the summer in which I fed upwards of 30,000 in one room, nobody was the worse for attending them; and yet I frequently spent whole days with them, as did many of those friends who were kindly attentive to assist me in the care of them.

I know a lady who had a good many silk-worms; she cleaned and fed them herself, and was seized with

with a bad fever. All this might be:—but without allowing for the coincidence of events, ſhe boldly aſſerts her fever to have been occaſioned by the ſilk-worms, and as loudly proclaims them unwholſome. It is, I ſuppoſe, from ſuch circumſtances as theſe, that the belief has gained ground; but I am decidedly of opinion that it is without other foundation.

The experiments you wiſh me to make, I certainly will attempt:—but I muſt confeſs the aſcertaining how much food will ſupport a given number of worms, has difficulties, ſince they eat much more voraciouſly at one time than another, and the lettuces vary materially in ſize.

There is a matter which appears to me of much greater magnitude, than the offering premiums for the planting of mulberry-trees; and that is, the holding forth a reward to thoſe who ſhall diſcover the beſt method to propagate them. All the gardeners with whom I have converſed on the ſubject are ignorant of the practice of any other way, than by tranſplanting the ſuckers which ſpring from the roots of the old tree; and theſe are ſo few in number, that the expence of the purchaſe muſt effectually deter any one from making a large plantation, whilſt the uſe and profits of it are ſo precarious.

Let

Let it once be known how they can be raiſed with eaſe, and in abundance, and the plan will become practicable, which it is not at preſent.

I am, Sir, your obliged friend,

HENRIETTA RHODES.

ARTICLE XLI.

The Deſcription and Uſe of Mr. Winter's *New-invented Patent Drill Machine.*

THIS Machine (ſays Mr. W.) is univerſally acknowledged to be ſuperior to any hitherto invented; it is ſo ſtrong that nothing but the greateſt violence can injure it; and is conſtructed on ſuch plain mathematical principles, as to be worked by any perſon of the loweſt capacity. It depoſits Grain, Pulſe, Turnip, Carrot, or any other Seed, with the greateſt accuracy, at any required depth in the earth, from the ſurface to ſix inches, at any required diſtance from 6, 7, 8, 9, to 40 inches between the rows, and may be inſtantaneouſly regulated ſo as to increaſe or decreaſe the quantity ſown, which is immediately covered. One man, a boy, and two horſes, can drill ten acres of light, and eight acres of ſtiff land in one day; and from one buſhel of ſeed wheat, and one buſhel and a half of barley, will produce a crop of from 6 to 20 buſhels per acre, (according to the richneſs of the ſoil) more than when ſowed by the common mode of huſbandry.

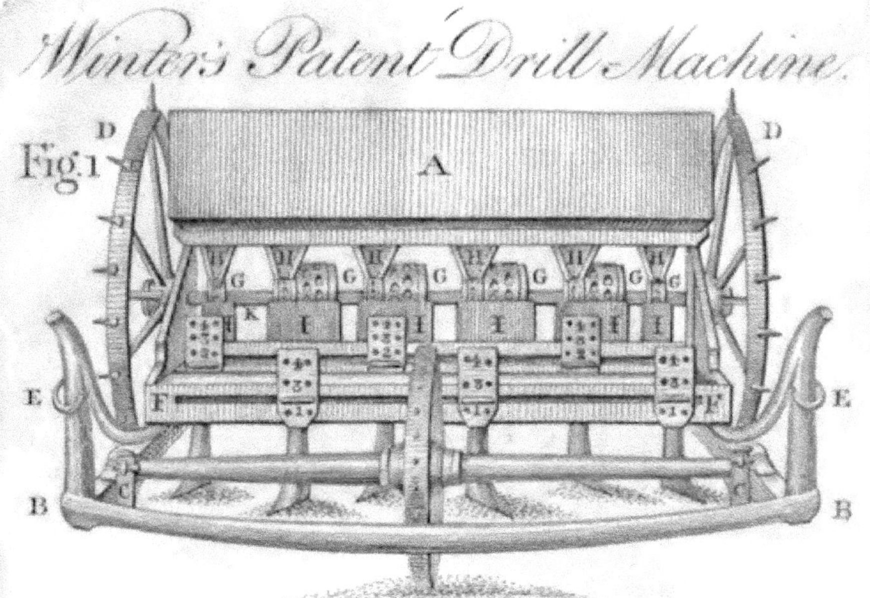
Winter's Patent Drill Machine.
Fig.1
A
D
D
E
E
B
B
F
F
K

Fig.2
Fig.3
d
a
g

The machine may be feen at the Exchange, or at Mr. HANCOCK's, wheelwright, Old-market, Briftol. Orders to be directed to Mr. G. WINTER, Briftol.—The price *Sixteen Guineas*.*

FIGURE I.

Reprefents a front view of the machine when at work, with fix coulters faftened on, depofiting grain at the depth of two inches, in drills at feven inches diftance.

A. The fore flap turned up, and the back board taken off, for the purpofe only of reprefenting the infide work, which when drilling in the field is all inclofed:—It then appears as a box between two wheels, and all the infide work is perfectly fecured from the effects of the moft tempeftuous weather.

B. The frame.

C. Iron plates, in which the gudgeons of the fore wheels are placed, and may be removed to any required depth.

D. D. The two hind wheels with fpikes, which are for the purpofe of preventing the wheels from fliding over rough ground or clods, and by the fpikes penetrating into the earth, the wheels are forced round, by which the grain is delivered; for when the wheels ftop, or flide, no grain is difcharged.

E. E. Iron rings faftened to the fore-ftandards, to which the chains are fixed, for drawing the machine.

F. Coulter-bars with grooves, through which the coulters are placed at any required diftance, from 6, 7, 8, 9, 10, to 36 inches or more.

* The faving of the feed and extra produce (more than can be obtained by the common mode of fowing) off ten acres of good land, drilled at the diftance of feven inches, with wheat at 5s. per bufhel, will in one feafon pay for the machine.

1, 2, 3, 4. Six Coulters numbered between the bolt-holes, with figures for setting the coulters so as to deposit the grain at any required depth; for instance, when the bolts are placed over No. 2, the grain is deposited at two inches deep; when over No. 3, at three inches deep; and so on.

G. Six cylinders, which occasionally slide off and on the axis, so that the whole, or any number of them, may be fixed at any required distance.

H. Boxes which contain the seed.

I. Conductors, into which the seed is delivered out of the cylinders, and conveyed into the grooves, in the back part of the coulters.

K. The axis, which passes through the cylinders and large wheels.

Figure II.

Represents a side view of the machine when at work.

a. Iron ring with a hook and chain fixed to it.

b. A sliding-board that covers an opening in the box, through which the axis and cylinders are taken out.

c. The case or box inclosing the works.

d. Handles for lifting up, and turning the machine at the headlands.

e. A marker for discovering the track of the machine, on land not ridged.

f. A harrow, for more effectually covering the seed and levelling the ground.

g. One of the three coulters on the fore coulter-bar.

h. One of the three coulters on the hind coulter-bar.

i. The guard which protects the conductors from being injured by stones, dirt, or weeds.

k. The

k. The pipe of the conductor, which enters into the cavity in the back part of the coulter.

There are two ftrong iron wheels which are placed in the back part of the frame, letter *e*, for the purpofe of travelling from one place to another; which wheels elevate the coulters about fix inches above the furface of the ground, and are immediately removed when fet to work. There are alfo regulators for increafing and decreafing the quantity of grain to be fown, which with the wheels cannot be difcovered in either of the views.

FIGURE III.†

Reprefents a running hoe for cutting the weeds between the drills, and adding earth to the rows of vegetables; the two points collect the weeds, which are in a manner inclofed, and more effectually deftroyed than when the blade is fquare, or angular.

G. W.

* Is not the invention of, but altered by Mr. Winter.

ARTICLE XLII.

No. 73, *Oxford-ftreet, London,*
March 20, 1778.

SIR,

AS it is but too common for individuals to fpeak roundly on fubjects in which they are particularly interefted; and as fuch affertions are generally viewed with a jealous eye by the public, I am induced to content myfelf with referring the public to individuals who have made experiments in

 drilling,

drilling, thereby putting myſelf out of the queſtion, and only obſerve, That from the rapid progreſs, which the drill ſyſtem has lately made, there is reaſon to apprehend that it will ſoon become general.

Your obedient humble ſervant,

JAMES COOKE.

Aſcertainments of Crops of Corn, reaped in the Year 1786; *the Seed of which was ſown by the Rev.* JAMES COOKE's *Patent Drill-Machine.*

Lord Viſcount Bateman, Shobdon, Herefordſhire. Wheat by the machine, 40 buſhels per acre. Barley, ſelf-evidently ſuperior to that ſown by hand, but omitted to be particularly aſcertained.

Marquis of Stafford, Trentham, Staffordſhire. Wheat by the machine, 33 buſhels per acre. Barley 24s. 6d. per acre more than broadcaſt.

Rev. H. J. Cloſe, Hill-Houſe, Ipſwich. Barley by the machine from 9 buſhels of ſeed, 400 buſhels of excellent grain. Alſo a particular experiment upon one quarter of an acre of poor land, and out of condition: by the machine 33 buſhels per acre; broadcaſt 19 buſhels. Oats, the produce in the ſame proportion.

Mr. Yeld, Milton, near Leominſter. Wheat by the machine 30 buſhels per acre; broadcaſt 25.

Mr. Boote, Atherſtone-upon-Stower, near Stratford-upon-Avon, the ſeven following accurate experiments, including in the whole 368 acres drilled, viz.—Wheat on loamy land, 47 buſhels per acre. On cold clay, 44 buſhels 5 gallons 2 quarts per acre. Wheat on cold

cold clay, 25 bushels 1 gallon 3 quarts. Broadcast on adjoining land, 9 bushels 4 gallons 2 quarts. Pease on light sandy land, 50 bushels 2 gallons per acre. Barley on light land 72 bushels 4 gallons per acre. Beans on light land, 36 bushels 2 gallons 2 quarts per acre.

Mr. Morley, Wood-hall, near Downham, Norfolk. Wheat by the machine, 44 bushels per acre.

Mr. John Lees, near Cirencester, Gloucestershire. Wheat by the machine, 53 bushels 4 gallons per acre. Broadcast, 39 bushels.

Moses Harper, esq; Astley, near Stourport. Barley by the machine, 56 bushels per acre. Broadcast 54 bushels.

N. B. The land of the hand sown crop allowed to be in better condition, than that upon which the machine was used.

Mr. Dunmore, Stanton-Wyvil, near Market-Harborough, Leicestershire. Wheat by the machine, 45 bushels per acre. Barley 72 bushels 4 gallons.

Mr. Glover, Burlaughton, near Shiffnal, Salop. Barley by the machine, from very light sandy land, 44 bushels per acre. Broadcast 37 bushels 4 gallons.

Mr. Hett, Bawtry, Yorkshire. Barley by the machine, 64 bushels per acre. Ditto broadcast, 48 bushels.

H. Cecil, esq; M. P. Hanbury-Hall, Worcestershire. Wheat by the machine, 5 bushels per acre more than broadcast, from two years' experiment.

Mr. R. Crabb, Moulton-park, near Northampton. Barley by the machine, 6 bushels per acre more than broadcast.

Colonel Wilson, Dedlington, near Stoke, Norfolk. Wheat by the machine, exactly half as much more as broadcast.

N. B. The drilled stubble, very clean by hoeing, the broadcast stubble a bed of poppies.

Sampson Barber, esq; Peterborough. Wheat by the machine, 27 quarters 6 bushels, from 5 acres 3 perches.

Mr. Wm. Wright, Warboys, near Huntingdon, an extraordinary crop of barley by the machine from fen land, allowed to be much ſuperior to broadcaſt.

Mr. Holland, near Louth, Lincolnſhire. Barley by the machine, 8 buſhels per acre more than broadcaſt.

Aſcertainments of Crops in 1787.

Sir Wm. Jones, bart. Ramſbury-Manor, Wilts. Wheat by the machine on a flinty loam, clover lay, one earth, 25 buſhels 1 gallon per acre. Ditto broadcaſt 20 buſhels 3 gallons. Wheat by the machine on a two-year clover lay, 27 buſhels 2 gallons per acre. Broadcaſt 25 buſhels. Barley by the machine after wheat, 27 buſhels per acre. Broadcaſt 22 buſhels.

Francis Skyrme, eſq; Lauhaden, near Haverfordweſt. Wheat by the machine, 57 buſhels 1 gallon per acre. Broadcaſt 48 buſhels 2 gallons. Barley by the machine, 67 buſhels 2 gallons per acre. Broadcaſt 48 buſhels 2 gallons. Oats by the machine 70 buſhels per acre: broadcaſt 49 buſhels.

Mr. John Boote, Atherſtone upon-Stower, near Stratford-upon-Avon. The eleven following aſcertainments, in all 450 acres, viz. —Beans drilled upon loamy ſand after oats, 50 buſhels per acre. Wheat drilled upon marl and mixed ſoil after beans, 36 buſhels per acre. Wheat drilled upon loamy ſand after beans, 50 buſhels 6 gallons. Barley drilled upon loamy ſand after turnips, 75 buſhels 5 gallons. Barley drilled upon ſandy land after turnips, 58 buſhels 4 gallons. Peaſe drilled upon loamy land after clover, 51 buſhels 4 gallons. Wheat drilled upon loamy land after beans, 45 buſhels 5 gallons. Oats drilled upon ſandy land after barley, 57 buſhels 3 gallons. Oats drilled upon loamy land after barley, 76 buſhels 6 gallons. Wheat drilled upon poor cold clay after clover, 25 buſhels 4 gallons. Wheat ſown broadcaſt upon poor cold clay after clover, on adjoining ground, 13 buſhels 7 gallons.

Mr. Boote has announced a clear profit of 700l. over and above his uſual profits, by drilling 450 acres in the year 1787. Alſo a clear profit of 500l. by drilling 368 acres in 1786.

Rev. H. J. Close, Dorking, Surry, so perfectly satisfied with his success in drilling, as to decline sowing any more broadcast. His estate in Surry, consisting of 700 acres, being now under the drill system.

Mr. Greenway Ascot, near Stratford-upon-Avon, Warwickshire. Barley by the machine, 16s. per acre more than broadcast.

Thomas Knight, esq; Godmersham-Park, Kent. Wheat by the machine, 4 bushels per acre more than broadcast.

Colonel Wilson, Didlington, Norfolk. Barley and oats by the drill, superior to any he ever had before.

Rev. J. S. Lushington, Bottisham, near Cambridge. Barley by the drill, 10 shocks per acre upon the field, more than broadcast.

M. Harper, jun. esq; Astley, near Stourport, Worcestershire. So perfectly satisfied with his success in drilling, as to decline sowing any more broadcast.

W. B. Earle, esq; Close, Salisbury. Eight acres of wheat by the drill, adjoining to 8 acres broadcast; the former superior to the latter, in the proportion of 8 to 7.——N. B. The 8 acres drilled with something more than 8 bushels of seed; the 8 acres sown broadcast, with 28 bushels.

Mr. John Austen, Old-Park, near Canterbury. Wheat and rye by the machine, infinitely superior to the broadcast upon adjoining land.

Mr. Taylor, Treasey-Farm, near Enstone, Oxfordshire. So far satisfied with his experiments by the machine, as to persevere in the practice.

Mr. Glover, Burlaughton, near Shifnal, Salop. The finest crop of pease by the drill he ever saw.

Mr. Quihampton, Repton, near Ashford, Kent. Wheat 4 bushels 7 gallons per acre more than broadcast.

Mr. John Stuart, and Mr. Jamett, Ashford, Kent. Wheat 20s. per acre in favour of the machine, compared with broadcast.

Nine

Nine other gentlemen, near Aſhford, expreſs themſelves perfectly ſatisfied with a ſuperiority in favour of the machine, without aſcertaining their crops.

Mr. Hall, Elmſtone, near Aſh, Kent. A comparative experiment in wheat, between Mr. Cooke's drill machine, and Mr. Ducket's ſyſtem of opening furrows in the land, and ſowing the ſeed broadcaſt. The reſult of the experiment was in favour of the drill-machine, which determined a wager of ten guineas in favour of Mr. Quihampton, of Repton.

To the above profits by drilling, may be added the average profit of 7s. or 8s. per acre of ſeed ſaved.

Article XLIII.

Deſcription of a new Harrow and Drag.

[Illuſtrated with an Engraving.]

Gentlemen,

I Take the liberty of ſending you a rough draft of a Harrow I lately invented, together with the motives which induced me to conſtruct one different from thoſe commonly in uſe; which when you have examined, compared, and proved, I doubt not but you will be induced to recommend to the publick,

At the time of ſowing barley laſt ſeaſon, I conſidered that our common harrows did not anſwer the end deſigned ſo well as could be wiſhed, and having

Pl. 2

Mr. Triffry's, Harrow & Drag

A Scale of Feet

having feen various kinds in different counties and places, I did not recollect to have feen any that anfwered better than my own. The faults I efpied in all thofe I had feen were, the tines or fpikes were placed too near together in the middle of the harrows, which prevented them from finking down into the ploughed land fo deep as neceffary; and when the ground on the top became fine, that they were apt to draw the couch, &c. together in heaps, and at the fame time little or no fervice was done for a foot or more within each fide corner; in confequence whereof we were obliged to double over that ground again, by which means I thought we loft at leaft one day's work in a week, which is no trifling matter.

After reflecting a little further on thefe things, I applied pencil to paper, and produced a plan, from which I had a harrow conftructed, which in feveral refpects far exceeded my expectations; the fides, or ends, do not only operate as well as the middle, fo as not to need doubling over again, but the tines or fpikes fink down fo much deeper than the common harrows, (on account of their being regularly placed at 15 inches diftance in each bar, whereby alfo the clods, &c. have a free paffage, and are not drawn together in heaps) but that no part is left untouched more than three inches, when the har-

row

row is drawn only once over the land; from all which conveniences, I find more execution is done by drawing this harrow once over the land, than any other I ever ſaw will do by being twice drawn over the ſame ſpot.

It equally excels as a drag, or firſt harrow, for rough land commonly ploughed; and alſo for ſuch as is turned one half on the other, which we call ſkirting or thwarting, as well as for finiſhing and ſmoothing the ſurface.

I imagined at firſt more ſtrength would be required to draw it, but find two of our little country horſes from 14 to 14½ hands high, draw it with eaſe.

My harrow is ſeven feet one inch long, and the poſts twelve inches from center to center. The bolts ſhould be forelock'd *on top*, and have a ſmall *flat head under*. The poſts are 3 by 2½ inches, the ledges or bars are three inches by 3-qrs. of an inch, which was intended for a finiſhing or laſt harrow, *not a drag*.

THE DRAG.

As to the drag, I obſerved the common one to ſcratch over the ſurface of the land without entering ſo deep as it ought, partly from the cauſe I before remarked

remarked in the common harrows, and partly from the tines being fixed ſtraight downward; but theſe being bent, and pointing forward, and alſo fixed at 18 inches diſtance, draw into the earth as deep as the ploughing, rending the ground in an extraordinary manner, and leave the *hard* clods to paſs freely through; yet no ground remains untouched more than three inches from the point of one tine to the point of another. The hinder poſt or rail is twice as heavy as either of the others, to keep down that part as deep as is neceſſary, which otherwiſe would not be the caſe. My land is in general tolerably free, ſo that four large, or ſix common oxen draw it well; but I prefer four of our little horſes to either.

I do not expect, that either the harrow or drag are ſo perfect as to admit of no improvement; but if they are inducements, only for ſome abler perſons to exerciſe their genius for the benefit of the public, I ſhall be amply rewarded for the thought, trouble, and expence, I have been at.

I am, with much reſpect, &c.

R. TREFFRY.

Beer-Barton, near Plymouth,
July 1787.

ARTICLE

Article XLIV.

On the Advantage of River Weeds as a *Manure.*

[In a Letter to the Secretary.]

Sir, *Norwich, Feb.* 29, 1788.

IN your Third Volume, is an insertion communicated by me, respecting River Weeds as a manure, when cut in their vegetating state, laid in the furrow, and ploughed in.

Permit me now to relate a mode of more experienced advantage, which is, by extracting them with their roots, and the surface of the soil on which they grow; leaving them awhile to the action of the sun and air, for a requisite fermentation; more particularly a certain species hereafter to be mentioned.

About the middle of June, in a broad part of a stream, where from a lessened current a muddy sediment rested, and on which, in spaces, various weeds grew, but whose surface was generally covered by the river Conferva,* whose extended deep green filaments scarce left any of the other species perceptible. To draw these out, I employed two men, accustomed to the cutting of weeds in rivers,

* Conserva rivularis of Linnæus.

who,

who, with paring and dragging inftruments, drew out many loads in the courfe of the day. Thefe, laid on a ridge, about fifty yards on the bank of the ftream, were continued there about three weeks; when I had two cart-loads of this aquatic fubftance laid on two different parts of an inclofed piece of land preparing for turnips, in an equal proportion with ftye and ftable manure, fpread at the fame time over the remainder of the field: with another load I filled up a hollow that had been lately excavated, on which I planted turnip-rooted and favoy cabbages; and at the fame time planted fome of both in a common garden foil, and likewife in fome unmixed mud, where no weeds grew, drawn from the river for that purpofe.

The virtue of the weed-compoft is obvious in each experiment; in the laft-mentioned, the favoys exceeded in cabbage, and the turnip-rooted in leaf and bulb, others of the fame fpecies fet in garden mould; while thofe fet on the mere mud have fcarcely made a progrefs; decifively evincing, I conceive, that the principle of increafe, and progrefs of vegetation, are more peculiarly derived from the weeds, than from the matrice on which they grew.

In refpect to the turnips, though no partiality was fhewn in fpreading more in quantity, in equal ſpaces, than of the other manures, nor were thoſe ſpaces in any eſtimated preference as to native ſoil; yet are thoſe ſpots diſtinguiſhed by a more vigorous vegetation, and a deeper green; nor can there be found on the reſt of the field (7-8ths of the whole) any roots ſo large as many in theſe ſmall tracks.

I have carried this experiment farther, but from thence no deciſion can be formed, being on land newly dibbled with wheat; the probable ſucceſs of which, and the more aſſured probability* of an improvement on a barley crop, intended in ſucceſſion to the turnips, I purpoſe, if in health, to communicate when time ſhall give the reſult; and am, in the interim,

Your very reſpectful friend,

JOHN WAGSTAFFE.

* To explain the apparent preſumption of more *aſſured probability*, it may be noticed, that of the river Conferva, many unbroken parts remained after the ſecond hoeing of the turnips, owing to the interwoven ſtate of its fibres, which are more immenſe than its extended leaves or filaments; and which, in the compound, before it was ſpread on the land, emitted a ſcent almoſt as ſtrong as ſtye manure.

ARTICLE

Article XLV.

An Account of a new Drill-Machine, invented by a Somerſetſhire Farmer, *and of a Crop ſown by it.*

[In a Letter to the Secretary.]

Sir, *Near Mells, Somerſet, March* 31, 1788.

IT is with pleaſure I comply with your requeſt, in giving you ſome account of the ſucceſs of my endeavours to complete a Drilling-Machine, which may unite the ſeveral objects of ſimplicity, general uſefulneſs, and cheapneſs. After conſiderable pains, I am of opinion, that I have ſucceeded ſo far as to inſure general ſatisfaction. And my own ſucceſs in the uſe of it, among my neighbours as well as on my own farm, confirms me very fully in the preference of the drill huſbandry before the broadcaſt, in all crops where the hoe is uſually introduced.

In the conſtruction of this machine, the peculiarities of which are entirely of my own invention, I have had regard to equal convenience for ſowing all ſorts of grain, and on ſoils and ſurfaces heretofore conſidered as the moſt unfavourable to drilling. I am about to procure a patent for the excluſive right of making this machine for ſale; for which reaſon, as well as that I have not procured an engraving of it, it, I omit attempting a particular defcription for the prefent. Whenever it fhall be fully before the public, I flatter myfelf it will be found to poffefs, at the price of Ten Guineas, at leaft more than all the valuable properties contained in other machines of a much higher price.

As I have now completed a new machine, including all the improvements fuggefted by confiderable practice with my firft; and as my wifh is to extend public utility, I fhall be happy to fhew it to any gentleman defirous of infpecting it. Moreover, as foon as I can get fufficient leifure, I intend making a model, and fending it to the Society's Rooms at Bath.

It may not be unimportant to fay, that this machine is conftructed to be drawn eafily by one horfe, fave on very rough and hilly ground. Having maturely confidered and proved the ufefulnefs of the different parts, I think myfelf warranted on the foundation of experience, and not of theory, in offering to fupply any perfon with this machine; having brought my workmen to fuch neatnefs of execution, as to bear a comparifon with moft others.

I intended fending you for the Fourth Volume of the Society's papers, an account of the quantity of acres fown laft year with the machine, for myfelf and

and neighbours, together with a ſtatement of the ſeed and crops; but find I ſhall not have time to do it correctly. I will, however, annex an account of a crop of peaſe ſown with my machine in its leſs perfect ſtate, and hoed with a large breaſt-hoe of my own invention.

In the beginning of April 1787, I drilled a field of 17 acres, of poor ſandy ground, on one earth, after barley, with eight ſacks and one buſhel of peaſe.* The rows were twice hoed, and earthed up with my breaſt-hoe before-mentioned, at the expence of leſs than 2s. 6d. per acre. My crop from the harveſt field was 63 or 64 full-ſized waggon loads; and though (not having completed my threſhing) I cannot exactly aſcertain the whole quantity of peaſe, I can do it ſufficiently near to aſſert, that I have above ten ſacks per acre, after deducting the ſeed, beſides the advantage of very clean ſtraw, and having my land in excellent order for my wheat crop.

With my breaſt-hoe, I found that one man could with eaſe hoe an acre per day, the firſt time of hoeing. And to ſhew the difference between this and the common hoe, I introduced three men with the

* This quantity, though but half what is uſually ſown broadcaſt, is greater than was neceſſary, and greater than I ſow this year.

latter to work in a field againſt two with the former. At night it appeared that the three had not more than finiſhed an acre, while the two with the breaſt-hoe had completed, in a much better manner, two acres. A ſtronger proof need not be brought to evince the ſuperiority of any inſtrument.

Brief Statement of the Advantage ariſing from my Management of the Pea Field above-mentioned.

DRILLED.		£. s. d.
Produce 170 ſacks, ſuppoſe at 16s. per ſack - - - - -		136 0 0
Seed 8¼ ſacks, at 18s. -	£.7 8 6	
Hoeing, twice, ſay at 2s. 6d. per acre - - -	2 2 6	
Extra expence on account of the ſuperior quantity to be brought to market, -	5 0 0	—14 11 0
Groſs profit -		£.121 9 0

BROADCAST.	
Suppoſe the ſame 17 acres had been ſown broadcaſt, the produce on the ſoil in queſtion would have been deemed good at 5 ſacks per acre, which at 16s. as above, would be - - -	£.68 0 0
Seed, 17 ſacks, at 18s. per ſack - -	15 6 0
Groſs profit on the broadcaſt	52 14 0
Difference of groſs profit on the 17 acres in favour of drilling - - -	68 15 0

or 4l. 0s. 10d. per acre.

It will be allowed, that my ſtatement of £.5 for extra expences of bringing the drilled crop to market is fully ſufficient, if not an unneceſſary abatement, when the quantity and cleanneſs of my ſtraw, and the ſuperior condition of my land, from twice hoeing, for a wheat crop, are taken into the general account.

I hope to give you ſome further account of my ſucceſs in ſimilar attempts; and though I employ the hand of a friend to methodize my communications, I deſire you and the public to conſider my veracity as pledged for matters of fact, and to be aſſured that I ſhall at any time have pleaſure in giving my opinion, and advancing the intereſts of huſbandry. Your's very reſpectfully,

GEORGE BARNES.

Article XLVI.

An Account of a Crop of Cabbages, for which a Premium of the Bath Society was awarded to the Writer.

[In a Letter to the Secretary.]

Sir, *Chilcompton, Dec.* 3, 1787.

AGREEABLE to your requeſt, I ſend you ſome account of my cabbage crop, of 12 acres, which the Committee has thought deſerving the premium offered by our Society. As you deſired,

I have weighed the produce of one perch, on each ſide of the road; there being, as you remember, ſome conſiderable difference in the appearance of the two pieces. The perch on that part which appeared lighteſt, weighed five hundred and one quarter; and ſixty times that weight, I believe, makes forty-two tons per acre,——The perch on the other ſide weighed eight hundred and a half, which is ſixty-eight tons per acre.——The former, as you juſtly obſerved, when they ſhall have attained their full growth, may be nearly equal to the latter. This difference I account for from the firſt having been ſown in the beginning of March, and therefore had not ſo good a chance for growth as the others, which were ſown in the autumn, and planted out in May. The ſpring-ſown ones were not planted out till near Midſummer, and then in ſo dry a time that they were almoſt ſcorched up. Therefore I ſhall in future join in opinion with my neighbour Mr. BILLINGSLEY, and always ſow for autumn plants. For which purpoſe, the beſt time to ſow the ſeed is about the middle of Auguſt, and tranſplant them off into ſome warm garden, or other place in which they may be ſheltered from very ſevere froſt.

In the next place I ſhall reply to your enquiry about the quality and general value of the arable land in this pariſh. It is of a light, ſhelly, ſtone-

braſh

braſh nature,—a ſoil in ſome people's opinion unfavourable to cabbage. They will tell you it ought to be a ſtiff clay, or heavy loam; my ſucceſs, however, in the growth of cabbages, proves that more is to be expected from manure and management than the diſputers about ſoils ſeem aware of. This remark, indeed, will apply to moſt crops, but you will pardon my making it.

The value of our arable land per acre, is about 30s. on the average; which is in my opinion too high a price to allow giving a direct ſummer fallow. But if the land were as low as half that price, I ſhould endeavour to raiſe turnips, cabbages, &c. as a fallow crop: ſuch a crop is worth more or leſs according to the price of hay, ſometimes 5l. 6l. 7l. or more, per acre, which is certainly at any rate an object, both to the farmer and the community at large; and far more eligible in moſt ſituations than letting the land lie for a bare fallow. For after the cabbage is fed off, (which I always endeavour to do, and ſow the land to wheat by Old Candlemas) I find, by more than ten years' experience, an additional advantage in the goodneſs of the following crop. Such wheat with me is ever ſuperior to that which I ſow at or before Michaelmas. The ſort of wheat I generally ſow after ſuch green crops, is

 the

the *white-eared*, so called at Warminster and Devizes; in the West, *brasil.*

I think you further wished to know the manure I made use of for my cabbage. It was a compost of lime, weeds, and earth, that lay under the hedges round the field, and a layer of dung, all mixed and turned together. I spread about 25 cart-loads on an acre, with the usual ploughing given to a common summer fallow. This is not to be reckoned with expences attending a cabbage crop; for admitting such crop to exhaust the manure in some degree by its growth, an ample restoration will be made by its refuse ploughed in, and by the stirring and cleaning the ground.——I will give you, as nearly as I can, a full account of the expences of the crop of cabbages per acre.

	£.	s.	d.
The seed sufficient for an acre, is ¼ lb. at 3s.	0	0	9
Sowing and transplanting - - -	0	5	0
Ridging up two furrows, and leaving two, with the Norfolk plough - -	0	2	6
Two men and two boys, for drawing and setting plants - - - -	0	3	4
Earthing up on the two furrows left, done when the plants are well rooted -	0	2	6
Hoeing and earthing up the said plants*	0	2	6
	£.0	16	7

* The plants ought to stand a yard apart every way.

	£.	s.	d.
	0	16	7
The carrying of the cabbages of the land to the cattle, as they are wanted, I will suppose, on such a scale of feeding as mine, to require a man, 2 horses, and a cart, half the day, which per week for one acre, is about - - - - - -	0	17	6
Total expence per acre -	£.1	14	1

The aforesaid man in my farm carries the cabbage to 45 oxen, and upwards of 60 sheep; and throws them out of the cart over the fields without cutting them. My 12 acres of cabbage will feed the above number of stock for three months, and I am very well assured that they prove as fast as they do in the prime months of the season, May, June, and July. I am, and would wish to be, the practical farmer; at the same time I am open to information, or the candid opinion of any man; remaining, dear Sir,

Your much obliged friend and servant,

HENRY VAGG.

[The communication of experiments conducted with care, and on so respectable a scale as Mr. VAGG's, above related, must be ever acceptable to this Society, as well as interesting to the publick. We would, however, beg leave to recommend to Mr. VAGG some future attention to the possible advantage of cutting the cabbages before they are eaten;

eaten; on which plan it may become prudent to contrive ſome mode of giving them to the cattle different from that of ſtrewing them on the ground. We are of opinion, that method and cleanlineſs in the fattening of every ſort of cattle, will ever be found of *ſome* importance, both in promoting their growth, and in the ſaving of their food.

Mr. VAGG has omitted to mention, and the preſs cannot now wait for the enquiry, what quantity of hay was eaten by his cattle, while feeding on the cabbage; but ſuch information, though a proper part of an experimental account, is not very important, as the quantity neceſſary to others may vary according to accidental circumſtances.]

ON THE
PRESERVATION OF THE HEALTH
OF
Persons employed in Agriculture,
AND ON THE
CURE OF THE DISEASES
INCIDENT TO THAT WAY OF LIFE.

BY

WILLIAM FALCONER, M.D. F.R.S.
And Physician to the Bath Hospital.

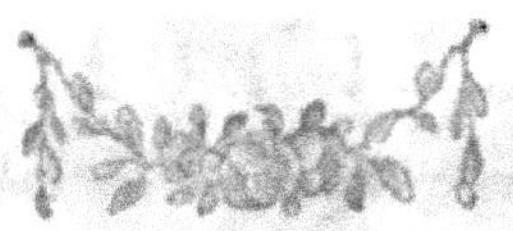

INTRODUCTION.

THE Preservation of the Health of persons employed in Agriculture, is, abstracted from moral and religious considerations, of greater national importance than any improvement either in the theory or practice of the art can lay claim to.

Without Artificers, it is obvious, that there can be no manufacture of any kind; and if the workmen are inferior in number to the proportion required, the business must languish, or be contracted in its extent. An attempt, therefore, to correct the errors, or to restrain the imprudence, with respect to such matters as concern Health, of those employed in this branch, can need no apology.

I wish I could say, that the execution of the work were equal to the importance of the subject; but however inferior it may be, it contains, I trust, some useful cautions, and, I hope, some hints that may lead to farther improvement.

I have judged it eligible, to address the present work to those who employ the persons for whose immediate use the cautions are principally intended, rather than to attempt to instruct the people themselves. Few of the latter have much time for reading, and little capacity for instruction in matters of reasoning. I have, therefore, ventured

tured

tured to direct myself to those from whom information of this kind would be likely to come with greatest authority, on account of their immediate connection and influence.

The Reverend Clergy will, I trust, excuse my offering a hint to them on this occasion. They are almost universally persons of liberal education, and more general knowledge, than falls to the lot of most of their neighbours. Would it not be an agreeable as well as an useful method of employing these advantages, to turn their thoughts towards the practical part of Medicine? The Natural History of the Human Body affords a more useful subject of investigation, than is done by stones, spiders, or shells; and though medicine, taken at large, is an arduous and deep study, yet it is practicable enough to gain sufficient knowledge of it to be of great service in many cases, especially such as occur most frequently among persons of the description here alluded to. The disorders incident to such are in general simple in their nature, and seldom exhibit at the same time such apparently contrary, and of course perplexing indications, as those which are the offspring of luxury and refinement.

All the disorders indeed that occur among such persons are not equally simple, but good sense and moderate information will suggest the propriety of asking, in such cases, the advice of persons whose professional education and attention has led them to a deeper knowledge of the subject.

ON

ON THE

PRESERVATION OF THE HEALTH OF PERSONS EMPLOYED IN AGRICULTURE.

ADVANTAGES WHICH PERSONS EMPLOYED IN AGRICULTURE POSSESS IN POINT OF HEALTH.

THE way of life of perſons engaged in agricultural buſineſs, exempts them from many of the diſorders to which other occupations are liable. Many of the employments by which great numbers of people are ſupported, are injurious to health, by being either too ſedentary, or too laborious; by which the powers of nature are either ſuffered to languiſh for want of exertion, or worn out prematurely by over-fatigue. But the buſineſs of huſbandry is not neceſſarily connected with either of theſe extremes. The labour is indeed conſtant, but not in general ſo violent as either to exhauſt the ſtrength by over-ſtraining, or to excite any weakening degree of diſcharge by perſpiration. The variety likewiſe of the neceſſary buſineſs is a favourable circumſtance for thoſe who are employed in it, it, as thereby the different muſcles of the body are exerciſed, and various poſtures uſed, which contribute to ſtrengthen the body more generally, and alſo relieve the mind by a diverſity of attentions.

A farther advantage attending the nature of labour in huſbandry is, that it is performed in the open air, which in general muſt be pure and wholſome, as being free from ſmoke and other vapours ariſing from inflamed bodies, and alſo from putrid exhalations both of the animal and vegetable kind, which are well known to taint the air in large cities, and in manufactories of every kind, where great numbers of people are aſſembled in a ſmall compaſs.

The ſurface or ſtaple of the ſoil, which is the ſubject of theſe operations, does not give out any noxious odours, like many of the mineral or metalline ſubſtances employed in ſeveral manufactories, but is at leaſt perfectly innocent, and has even been thought to produce effluvia rather favourable than injurious to health. The number of vegetables, likewiſe, with which perſons concerned in ſuch employments are generally ſurrounded, contribute to render the air which is reſpired pure and ſalubrious, by abſorbing the putrid and phlogiſtic ſubſtances that float in the atmoſphere.

The

The diet of perſons who live in the country is, I think, in general more wholſome than that of thoſe who inhabit towns. A large portion of it conſiſts of freſh vegetables and milk, which, though not excluded from the food of thoſe who live in towns, are enjoyed in much greater plenty and higher perfection in rural ſituations. Theſe correct the putrefactive diſpoſition of animal food, and tend to keep up the proper ſecretions and evacuations, and to maintain that balance in the animal ſyſtem, upon which health ſo much depends.

The regular hours neceſſary to be obſerved by thoſe who follow country buſineſs, are perhaps of more conſequence than any of the other articles, however important thoſe may be.

It is an old and a common opinion, that the external air is much leſs ſalubrious during the night than the day; and this opinion, which probably was at firſt drawn from obſervation, ſeems to be confirmed by chemical experiments, which tend to ſhew that the air exhaled by vegetables, whilſt the ſun is above the horizon, is much more pure and fit for reſpiration than that which iſſues from them in the abſence of the ſun. The ill effects of the latter are probably beſt avoided, by the human body being in a ſtate of repoſe and inſenſibility, which render it leſs liable to be affected by ſuch impreſſions. The morning air, on the contrary, ſo celebrated both by poets and philoſophers for its benign and cheering effects upon the mind and body, is enjoyed in high perfection by perſons in this way of life: and the advantages they derive from thence in point of health are probably very great.

I have been informed from the beſt authority, that a perſon in high ſtation ſome years ago, who was very deſirous to protract his exiſtence in this world as long as he was able, made every poſſible enquiry concerning the regimen and manner of life of thoſe perſons who had arrived at a great age, but found no circumſtance common to them all, ſave that they all had obſerved great regularity in point of hours; both riſing early, and going early to reſt.

Freedom from care and anxiety of mind is a bleſſing, which I apprehend ſuch people enjoy in higher perfection than moſt others, and is of the utmoſt conſequence. Mental agitations and eating cares are more injurious to health, and deſtructive of life, than is commonly imagined; and could their effects be collected, would make no inconſiderable figure in the bills of mortality.

The ſimplicity and uniformity of rural occupations, and their inceſſant practice, preclude many anxieties and agitations of hope and fear, to which employments of a more precarious and casual nature are subject. Nor is it the least advantage to health, accruing from such a way of life, that it exposes those who follow it to fewer temptations to vice than persons who live in crouded society. The accumulation of numbers always augments in some measure moral corruption, and the consequences to health of the various vices incident thereto, are well known.

Disorders to which Agricultural Persons are subject from the Nature of their Employment.

THE life of husbandmen and farmers, though in general healthy, has, like other situations, some circumstances attending it which produce disorders. These may be considered in several points of view, according to their causes.

First, then, the nature of their employment often exposes such persons to the vicissitudes of weather. These, perhaps, may be of many very different kinds, when considered with regard to the changes n the nature of the atmosphere; but this is an enquiry too deep and obscure for a popular treatise, ike the present, and I shall only take notice of such as are obvious and certain. These are three in number, *cold*, *heat*, and *moiſture*; to which may be added, a combination of the laſt of theſe with either of the former.

Expoſure to a great degree of cold may produce inflammatory diſorders of different ſorts, but principally, though not altogether, of the topical kind. Thus the inflammatory ſore throat, rheumatic pains in the teeth and face, inflammations of the eyes, and coughs, with pain of the breaſt, attended with fever, are all complaints liable to be produced by cold air, either externally applied, or drawn in by the breath. To theſe may be added, the rheumatiſm, both of the acute and chronic kind, which, though ſometimes a local diſorder, is often general, and may be frequently traced to this cauſe.

Cold, likewiſe, when great, and long continued, is apt to produce diſorders of an oppoſite nature to thoſe juſt mentioned. Paralytic affections are frequently cauſed by it, eſpecially in the lower extremities, which are generally the moſt expoſed to its influence.

Heat is another ſource of diſeaſe to the huſbandman, who often experiences its bad effects in time of harveſt. Inflammatory fevers are often the conſequence of heat and labour, and ſometimes ſuch as are attended with local inflammation, as pleurisies, peripneumonies, inflammations of the bowels, &c. Sometimes the brain is primarily affected, probably from the immediate effect of the sun's rays upon the head. The eyes are also liable to be inflamed from exposure to strong light. Moisture, especially when combined with either of the above extremes of temperature, is productive of several disorders.

People who work in the open air, and oftentimes at a distance from shelter, must necessarily be exposed to casual showers at every season of the year. If these happen in cold weather, they aggravate the bad effects of cold, by conveying it to a closer contact with the skin, and also by the generation of cold by evaporation. If rain fall suddenly at a warm season of the year, its effects are, I apprehend, less dangerous than in cold weather to those who are wet with it; nevertheless it is not void of hazard, especially if the persons exposed to it have been previously much heated, either by the weather, or exercise.

The evaporation of the moisture generates a degree of cold, which is greater as the evaporation is quicker. This then is one reason, why the danger of wet clothes is greater, as the body is more heated.

Whether moiſture, ſimply conſidered, has any other effect than as increaſing the influence of cold, is not clearly determined. But whatever doubts we may entertain, concerning the moiſture of the atmoſphere, there is no queſtion that ſome kinds of moiſture, to which perſons who labour in this way are ſometimes expoſed, has ſpecifically noxious qualities.

The draining of marſhy grounds, however it may in its conſequences benefit the health of thoſe who live in the neighbourhood, has been long obſerved to be but an unwholſome employment for thoſe who work at it. Yet this is frequently a neceſſary piece of buſineſs for the farmer, as well as the cleanſing of ditches, which is in ſome meaſure of the ſame kind, though in general leſs apt to do miſchief. The moiſture to which people thus employed are expoſed, muſt not be conſidered as mere humidity; but as humidity, combined with putrefying ſubſtances, and capable of diffuſing the effects of ſuch over thoſe who are within a certain diſtance of it.

Marſhes are well known to produce diſorders, even over a conſiderable extent of country, and muſt of courſe be particularly liable to affect thoſe who break up any part of them. Putrid complaints of various kinds may be produced by theſe exhalations;

exhalations; but I apprehend, the intermittent fever is the uſual conſequence; the frequent appearance of which, in moiſt and fenny countries, has been univerſally obſerved.

Such are the diſeaſes to which people employed in huſbandry are occaſionally liable from the nature of their occupation. But they are ſubject to a much greater number from their own imprudence, of which I ſhall next ſpeak.

Diſorders to which Perſons employed in Agriculture are liable from their own imprudence.

AND firſt, *Of their wanton expoſure of themſelves to the viciſſitudes of Heat and Cold.* It is no uncommon thing for people who work in harveſt, when violently heated by the weather and by labour, to drink large draughts of ſome cold thin liquor, as water, milk, whey, butter-milk, and ſuch like. This, if taken in great quantity, has been ſometimes known to ſuppreſs the powers of life altogether, and to produce an almoſt inſtant death.

This however, I believe, ſeldom happens; but the bad effects of this practice appear in other ways fufficiently ferious to difcourage fuch hazardous experiments. It is not uncommon for a violent fever to be the confequence, which is frequently attended with inflammation of the ftomach or bowels; both which are diforders of the moft dangerous nature. But fhould they efcape incurring any acute complaint, it is common for them to be affected with a fenfe of weight and ficknefs at the ftomach, which continues feveral weeks, and is at laft relieved by vomiting; this, however, does not put a period to the complaint, as it is generally followed by an itching eruption on the fkin in blotches, in various parts of the body, which proves to be the leprofy—a loathfome and filthy difeafe, and very difficult of cure!

I have had an opportunity of feeing at the Bath Hofpital, a great number of people thus afflicted, and am fatisfied that they all, without exception, owed their difeafe to the application of cold, in fome form or other, to the body when in a heated ftate.

Labouring perfons are very apt, when they leave off any work in which they have been much heated, to remain fome time at reft in the open air before they put on their clothes. This is a very imprudent practice, and frequently produces bad effects, efpecially in bringing on coughs, and other diforders of the breaſt, which oftener owe their riſe among the common people to this than any other cauſe.

Neglect of changing their clothes when wet, is alſo a great ſource of diſorder among huſbandmen. To remain in wet clothes when the body is at reſt, ſubjects the perſon who is ſo imprudent as to ſuffer it, to the united bad effects of cold and moiſture. Much worſe conſequences may however be expected, when they who are heated by labour lie down to ſleep, as they often do, in their wet clothes. The diminution of the force of the circulation and other powers of life, which always takes place during ſleep, cauſes the bad effects of cold to operate with much greater danger to health and life. This hazard is much aggravated, if they add to this imprudence by ſleeping on the wet ground. This not only communicates an additional moiſture and cold, but is perhaps ſtill more prejudicial from the nature of the exhalation. It is the opinion of a phyſician of the greateſt eminence, that the vapour which ariſes from moiſt earth is the cauſe of the moſt dangerous fevers. Thoſe, therefore, who put themſelves wantonly in the way of ſuch danger, are guilty of little leſs than ſuicide.

Exceſs, or *Irregularity in Diet*, is another ſource of diſorder to people in this way of life. This is common indeed in ſome meaſure to all ranks, but in ſeveral reſpects it is particularly applicable to thoſe who are employed in huſbandry. Air and exerciſe are well known to ſharpen the appetite; and as theſe advantages are incident to this way of life, it may be expected that ſome exceſs ſhould now and then take place. The diet of ſuch perſons is indeed in general too ſpare and plain to offer any great incentive to indulgence in point of quantity, but opportunities ſometimes offer for a more plentiful allowance of food, and more inviting to the palate. On ſuch occaſions the lower ranks of people exert little conſideration or prudence. They have ſcarcely any view beyond the gratification of the preſent moment; and if a full indulgence of appetite is not exerciſed, they deem it a loſs of an opportunity for the enjoyment of ſo much happineſs.

It is needleſs to enumerate in this place all the complaints that exceſs in quantity of food may bring on; it is ſufficient to ſay, that it has often produced ſudden death, and where its violent effects have not been ſo immediate, has laid a foundation for bad health during the remainder of life.—To this head may be referred the brutal practice of eating enormous quantities for a wager, or out of bravado. It is needleſs to deſcant upon ſo odious a ſubject, farther than to ſay, that ſuch things ſink

men

men below the level of beafts in groffnefs and folly, not to mention the fcandalous immorality of fuch actions.

The diet of people employed in hufbandry, does not admit of much luxury refpecting its quality; there are however fome things which come within the reach of thefe people, and which they regard as gratifications, and of courfe are apt to take in too great quantity. Of this kind are fome of the autumnal fruits, which in fome years are produced fo largely, as to be of fcarcely any pecuniary value. Of thefe, plumbs, efpecially fuch as are of the coarfer and more auftere forts, are the principal. It is a common obfervation, that in years wherein there is an abundance of fuch fruits, purgings, colicks, and moft other complaints of the ftomach and bowels, are very common. It is proper here to obferve, that the incautious manner in which thefe fruits are devoured, efpecially at their firft coming in, caufes many of the ftones to be fwallowed;—a practice extremely hazardous. The hiftory of phyfick affords many examples of the worft confequences arifing from fuch bodies lodging in the ftomach and bowels. Sometimes, when the accumulation of them has been confiderable, they have obftructed the alimentary canal altogether, and produced a miferable death in a fhort time; at

others,

others, they have made their way through different parts of the body, and caused either a long and painful illness, or death, by the hectick fever attending internal suppurations.

Pears, if eaten too freely, are apt, as well as the stone-fruits, to disorder the stomach and bowels; but they are less dangerous, and not so often swallowed in such quantities as to be materially prejudicial to life or health.

Nuts are perhaps, upon the whole, the most dangerous of any of the fruits that are likely to fall into the way of this rank of people. When eaten in large quantity, they have been often known to lodge in the stomach, and to be incapable of being removed from thence by any medicine, and of consequence have put a speedy end to life. When taken in less quantity, they are found to oppress the breathing, and to produce vomiting and bowel complaints.——Hoffman observes, that dysenteric complaints are always most common in those years in which the harvest of nuts is plentiful. Excess in diet, however, is more frequently committed in liquids than in solids.

It is observed of mankind in general, that they have a natural fondness for fermented or spirituous liquors,

liquors, and a certain proportion appears to be allowable and even neceſſary for perſons who undergo hard labour. But the healthy quantity is apt to be exceeded when opportunity offers, and exceſs of this kind is more hurtful than a defect of ſuch gratifications. I need not here enlarge on the conſequence of *drunkenneſs* to health. Fevers, dropſies, conſumptions, apoplexies, and many other miſerable diſorders, are well known to follow ſuch a courſe. The want of money among labouring people, indeed often prevents the bad effects of a habit of this kind, but occaſional opportunities occur which are laid hold on with great avidity; and it is far from uncommon to find death the immediate follower of ſuch licentious indulgence.

Diet, however, is not the only article which ſuch perſons are liable to carry to exceſs. It is common to ſee exertions of a more liberal kind purſued to too great length. The caprice of emulation will often produce inſtances of labour, which duty, and the urgency of circumſtances, might in vain ſolicit. The burſting of ſome of the blood-veſſels, particularly thoſe of the head, lungs, or ſtomach, nephritic complaints, and inteſtinal ruptures, have all of them followed ſuch ill-judged and oſtentatious diſplay of ſtrength and corporeal abilities.

Directions relative to the Prevention and Cure of Disorders incident to Persons employed in Agriculture.

AFTER the above enumeration of complaints to which persons thus employed are liable, it is proper I should offer something on the subject of their cure or relief. This I shall consider in two views; the first as to what regards the prevention of disorders, and the second as to what regards their cure.

Persons that work in husbandry are necessarily exposed to the weather in both its extremes of temperature. The ill effects therefore of both, it behoves us to counteract. Cold in this climate is most necessary to be attended to, as its operation is of longer duration; several months in the year often requiring us to be on our guard against cold, whilst excessive heat scarcely lasts more than a few days. Warmth of clothing is the only method, exercise excepted, by which those who spend their life in the open air can guard against cold, and nothing is more necessary for such persons as are the subjects of the present consideration, than a proper regard to this article.

The woollen cloths of our own country are perfectly well adapted for these purposes, being warm without being too heavy, resisting moisture in a good measure,

measure, and even when wetted being less cold to the touch than any other substance. It appears to me that some of the coarser and looser woven fabricks are preferable, both in point of warmth and lightness, to those of a more even surface, and also give more resistance to the penetration of moisture.

Every person who employs men under him in business of this nature, ought to be careful, in point of interest as well as humanity, that his servants have clothing sufficient for the season of the year; otherwise he may expect a proportionable diminution in the labour he expects to be performed, and the loss of many valuable opportunities, especially in precarious weather and seasons.

The same arguments are applicable to those who have the care of the parish poor, whom it would be far more œconomical as well as humane to preserve in a good state of health, than to suffer them to become victims of diseases which might be prevented. This caution refers particularly to the youth, who, by being neglected at that time of life, often continue burdens on those persons, whose expences (had the children's health been duly attended to) they might have contributed to diminish.

Friction, properly applied, might prove an excellent preservative against, and even a remedy for

many

many of the bad effects of cold. Would persons chilled with the severity of the weather, rub their bare limbs with woollen cloths for a considerable time after they return home, it would produce a more equable and genial warmth, and contribute more to support the powers of life, than any artificial heat whatsoever. The same operation would probably prevent many of those painful and refractory sores called chilblains, which are so apt to affect the extremities, especially in young people. Should any persons in extreme frost have their limbs or any part of the body actually frozen, the utmost caution must be had not to bring them near to any fire. The safest method is said to be, to rub the part frozen first with snow, and to continue the friction till some degree of warmth begins to appear, but not to suffer the access of any heat from fire, till the warmth from friction takes place. Even then, the part frozen should not be suddenly exposed to the heat of a fire, but rather be continued to be rubbed till the natural sensation and heat are perfectly restored. If the part frozen be exposed to the heat of a fire whilst in a frozen state, it will undoubtedly mortify.

It seldom happens, that the cold is so intense in this country, as to destroy those exposed to its influence by its direct and immediate operation; yet

as

as great degrees of it now and then take place, it may be proper to caution thoſe who may be in a ſituation that expoſes them for any conſiderable time together to violent cold, to be cautious how they ſuffer any propenſity to ſleep, or drowſineſs, to ſteal upon them. A tendency to ſleep in a perſon who is in ſuch a ſtate, is a certain ſign that the cold begins to gain ground on the powers of life, and ſhould therefore excite the ſtrongeſt efforts to reſiſt it. This may be a difficult taſk, but is neceſſary, as life entirely depends upon it.

Heat, though leſs frequently an object of our care in this reſpect than cold, nevertheleſs demands our attention. Though ſeldom of long duration, the heat is ſometimes exceſſive. I have ſeen it in the ſhade, and in a ſituation expoſed to no reflected heat, raiſe the thermometer to 87 degrees. Such heats, and even conſiderably leſs, are too great for laborious work even in the ſhade, and muſt be ſtill more injurious to thoſe who are expoſed to the ſun's rays, which is of neceſſity the caſe with thoſe who work in the harveſt.

In ſuch extremities of temperature, it ſhould not be expected, or even permitted, that the unthinking labourer, who has ſcarcely any views beyond the preſent moment, ſhould expoſe himſelf to ſuch ha-

zard.

zard. Œconomy, as well as humanity, pleads loudly in behalf of ſuch indulgence.

Inferior, yet ſtill conſiderable degrees of heat, although they need not preclude work in the open air, ſtill have need of ſome cautions reſpecting them. It is not uncommon to obſerve a degree of impatient anxiety which accompanies ſome people in every action of life. This prevails among the lower as well as higher ranks of mankind, and often proves a ſource of fatigue and toil, without expediting labour. Calmneſs and compoſure are neceſſary to the corporeal as well as the mental operations, and tend greatly to prevent the bad effects of exceſs of ſtimulus of any kind.

As the head is the part principally expoſed to the action of the ſolar rays, it is particularly neceſſary to uſe ſome defence for that part. Hats are uſed for this purpoſe, but the black colour of which they are generally made, cauſes them to abſorb the heat, and of conſequence to accumulate it in the very part on which we ſhould leaſt deſire it to fall. Hats for working people in hot weather ſhould be made of ſtraw, or ſome light ſubſtance of a white or pale colour, and with brims ſufficiently wide to ſhelter both the head and ſhoulders from the ſcorching beams of the ſun. Even a piece of white paper

covering

covering a hat, is no contemptible defence againſt ſolar heat.——The eyes ſhould likewiſe be conſidered, which expoſure to ſtrong light is ſo apt to injure. This ſhould be guarded againſt by the brim of the hat being made of a ſufficient breadth to ſhade the eyes, and the inſide ſhould alſo be tinged of either a green or blue colour, but by no means either black or a very light hue.

I have before mentioned the bad effects of cold applied in any way to the body when violently heated. This ſhould ſerve as a ſufficient caution againſt ſuch imprudencies. It may be uſeful to add, that as it may be neceſſary to drink frequently, it prevents much of the bad effects of cold liquor, to eat ſomething ſolid immediately before any liquid be taken. A few morſels may be ſufficient, and the efficacy of the precaution is well known.

The miſchievous conſequences of cold liquors, drunk in ſuch caſes, are much aggravated when they are, as is too common, ſwilled down in enormous draughts. Would thirſty people but have a little patience, and drink ſmall quantities at a time, with proper intervals, as of a few minutes, the uneaſy ſenſation would be more effectually removed, and that without any danger to health.

Another caution highly neceſſary for ſuch perſons is, to put on their clothes immediately on their leaving off work, and to do this without any regard to the warmth of the weather. Nothing can be more hazardous than for a perſon who is heated with labour, and in a ſtrong perſpiration, to remain expoſed to the wind. The exhalation both from the body and the wet linen, produces a ſudden and conſiderable degree of cold, which is not merely tranſient, but continues as long as the moiſture is ſuffered to exhale freely into the open air.

I have before remarked the hazard of labouring perſons ſleeping on the ground during the intervals of their work. This is improper at all times, but particularly dangerous if the ground be any wiſe moiſt. Indeed I am of opinion that ſleep had better be avoided altogether at ſuch times; as ſuch ſlumbers produce but little refreſhment, and expoſe the health to unneceſſary riſque. The body would be ſufficiently reſted by the ceſſation of labour, and early hours in the evening would afford a ſufficient portion of time to be ſpent in ſleep.

Moiſture is equally neceſſary to be conſidered in this place, with reſpect to its effects on the health, as *heat* and *cold*. This, I have before obſerved, cannot be always avoided, but the bad effects it

ſometimes

ſometimes produces may generally be obviated. If thoſe who are wet with ſhowers, would be careful to continue their motion and labour whilſt they remain in the open air, and to change their clothes on their return home, many of the bad conſequences of wet clothes would be prevented. Friction on ſuch occaſions might be an excellent preſervative againſt the bad effects of cold and moiſture: were the body and extremities that have been ſo expoſed, rubbed ſtrongly for a quarter of an hour with a coarſe woollen or linen cloth, immediately on the wet clothes being ſtripped off, it is probable few bad conſequences would follow from the accident.

It is indeed extraordinary this ſhould not be oftener practiſed in ſuch circumſtances than it is. Every labouring man knows the neceſſity of rubbing horſes that have been wet and dirty, and this not only for the purpoſe of cleanſing away the filth, but alſo for that of preſerving a due perſpiration and regular warmth on the ſurface of the body. Bathing the feet in warm weather would alſo be an uſeful precaution on ſuch occaſions, eſpecially to thoſe who are ſubject to purging and other diſorders of the bowels.

Labouring men are ſometimes expoſed to moiſture of a leſs innocent kind than ſuch as falls from the clouds. Draining marſhy ground is a neceſſary buſineſs, and, as I have before ſaid, expoſes the workmen to hazard from the nature of the moiſture, as well as from ſimple humidity. The intermittent fever is the principal, though not the only complaint, work of this kind is liable to bring on, and muſt be particularly guarded againſt. It therefore ſeems proper that ſuch kind of work ſhould, if poſſible, be performed in the ſpring, or early in the ſummer, in which ſeaſons theſe diſorders are not ſo likely to happen as when the autumn is advanced. And thoſe who work in this way ſhould be ſufficiently clothed, and be very cautious to avoid ſudden tranſitions from heat to cold.

Intemperance is particularly dangerous under ſuch circumſtances. It is highly proper, and even neceſſary, that thoſe who perform ſuch kind of labour ſhould have a ſufficient, and even liberal allowance, in point of diet; but exceſs of any kind, in ſpirituous liquors eſpecially, tends to weaken the ſtomach, and in conſequence thereof, the whole vital ſyſtem, and to render the body more liable to receive contagion of every kind. This is not a caution founded merely on theory or general principles, but a fact in medicine eſtabliſhed beyond all doubt.——Another caution very neceſſary to be attended to is, that none ſhould go to ſuch kind of labour in the morning

ing before they have taken ſome kind of food. Somewhat warm is moſt proper, and if it can be had, I ſhould prefer animal food. It is difficult to account for, but true as a fact, that warm victuals are greatly more cordial and ſtrengthening to the body, and of courſe more fit for the ſupport of thoſe who perform laborious work, than the ſame food if taken when cold.

Cleanlineſs is an eſſential article in ſuch circumſtances. Would thoſe who work at ſuch employments be careful to waſh their hands and feet at their return from work, and to change their linen and ſtockings as often as their circumſtances would admit, it is probable that the hazard would be greatly leſſened.

It is neceſſary to remark, that the above cautions apply at leaſt equally ſtrong to thoſe who ſuperintend ſuch operations, as to thoſe who actually perform them. It is probable that the labour of body and attention of mind, which occupy thoſe who are at work, is no ſmall preſervative againſt the acceſs of contagion of every kind.

As there is reaſon to believe, that intermittent fevers may in ſome caſes be ſo far infectious as to be communicated from one perſon to another, it would be proper that when any perſon ſhould be attacked therewith, ſuch perſon ſhould be provided with a ſeparate bed during the continuance of his diſorder. Cautions of this kind would be the beſt œconomy, as diſorders might then be checked at their firſt appearance, and prevented from ſpreading.

Exceſs, or *Irregularity in Diet*, is the next ſubject of theſe cautions. I have before mentioned ſome of the diſorders likely to be produced hereby, but ſhall now be more particular. Food may be conſidered with reſpect to its *quantity* and its *quality*. The firſt of theſe can only be meaſured by a reaſonable attention to the appetite. What may no more than ſuffice for one man, may be great exceſs in another; and in general what the appetite leads to, may be conſidered as the proper ſtandard. But ſome ignorant ruſtics are fooliſh enough to imagine, that there is a degree of credit annexed to the being able to conſume a larger quantity of victuals than is in the power of other men; and this beaſtly prejudice, which often produces fatal conſequences, ſhould be as much as poſſible diſcouraged, even by thoſe who practiſe hoſpitality among the lower ranks of people. It is certainly mean to offer to entertain any perſons, of whatſoever degree they may be, without producing a ſufficient quantity of wholſome proviſions;

viſions; but it is ſtill more inhoſpitable to encourage any to make ſuch an uſe of what is provided for them as to endanger health or life, not to mention the ſcandalous waſte which muſt be cauſed by it. Still more blameable is the practice of encouraging gluttony by wagers, or offers of reward. They who do this are in fact highly criminal, and in no ſmall degree guilty of the fatal conſequences which ſo often follow ſuch brutal diſplays of appetite. Moderation is not only neceſſary in what regards the quantity of food, but alſo as it regards the time in which it is conſumed. It is neceſſary to the proper digeſtion of our food, and of courſe to the nouriſhment of the body, that it be taken in gradually, and its texture broken down by chewing. It would ſcarcely be credited, were it not known as a fact, that the folly of gluttony has prompted wagers not only on the quantity of food, but alſo on the time in which it ſhould be ſwallowed; by accelerating which, all the bad effects of an enormous quantity of victuals muſt be greatly aggravated. Meat, thus ſwallowed, muſt be of courſe in large pieces, ſcarcely acted on by the teeth, and of difficult digeſtion. The ſudden diſtention of the ſtomach, by the introduction of a large quantity of meat ſo nearly at the ſame time, muſt weaken its tone, by overſtretching its fibres; and this has ſome-

times

times gone to ſuch a length as to deprive the ſtomach of all that power of expelling its contents, which ſoon terminated in death.—To theſe dangers ſhould be added, that of the meat ſticking in the paſſage of the gullet, and remaining there without a poſſibility of removal, a thing which is not uncommon amidſt ſuch exceſſes. Even the proper temperature of food is worthy attention. Ruſtick folly has produced wagers and premiums on the eating food nearly boiling hot. It is difficult to preſerve any temper in the cenſure of ſuch outrageous ſtupidity.

The quality of food is neceſſary to be conſidered, as well as its quantity. The ſtomachs of labouring men are undoubtedly ſtrong, and able to digeſt coarſe meat; but ill-judged œconomy ſhould not prompt farmers to ſet before their ſervants decayed or indigeſtible food. Putrid meat is not merely unwholſome in its remote conſequences, but immediately dangerous to life, as has been often experienced, and ſhould be avoided as carefully as we would any other poiſonous ſubſtances.

I have before pointed out the bad conſequences that are apt to reſult from the free uſe of ſome indigeſtible fruits, particularly plumbs and nuts. I think it would be a proper caution for farmers not to plant any of the former that are of the coarſe and auſtere kind; for though they generally bear plentifully, their fruit is of little value, and likely on that account to fall to the ſhare of ſuch people.——Hazel plantations are more neceſſary; but ſtill it would be of ſervice to place them as far from farm-houſes as might be convenient, that they might afford leſs temptation for the gathering of their fruit. It is proper to notice here the danger thoughtleſs people who ſpend much time in the fields are expoſed to, from eating plants and berries with which they are unacquainted. Many plants, commonly met with, are well known to be extremely poiſonous, ſuch as the Henbane, Deadly-nightſhade, Water-hemlock, ſome ſpecies of Drop-wort, ſeveral kind of Muſhrooms, and many others.——It ſhould be a ſtrict injunction to all who ſpend their time in the fields, never to taſte any plant, fruit, or berry, which they do not know to be ſafe, and indeed it would be more prudent to diſcourage altogether ſuch uſeleſs curioſity. It is obvious that this caution is particularly neceſſary for children.

The danger of exceſs in liquids is greater than in ſolid food. Fermented liquors, taken in moderate quantity, are both proper and neceſſary for thoſe who perform laborious work; but this healthy proportion is apt, when opportunity offers, to be exceeded by people whoſe gratifications are few in

number,

number, and of rare occurrence. As it is impracticable to prevent ſuch exceſſes altogether, I would wiſh to ſuggeſt, that, if they muſt take place, malt liquor is found by experience to be much leſs injurious to the health and conſtitution than diſtilled ſpirits, however diluted with water. I have been informed that a principle of œconomy has induced many farmers to treat their ſervants, and thoſe with whom they are connected, with ſpirits and water, inſtead of malt liquor; but ſuch a practice is by all means to be diſcouraged, as ſpirits are much more inflammatory than malt drinks, and produce more readily obſtructions and inflammatory diſorders, eſpecially of the liver and meſentery. The temporary delirium of intoxication that they produce, is ſaid to be much more violent and outrageous, and of courſe more dangerous, than what follows from taking too large a quantity of malt liquor. The deſtructive effects of ſpirituous liquors were ſo obſervable ſome years ago, as to produce the moſt ſerious apprehenſions in a national view, and to attract the notice of the legiſlature.—The baptiſms of London alone are ſaid to have been reduced from twenty thouſand annually to fourteen thouſand, which was with reaſon aſcribed to the uſe of this pernicious beverage:—This fact is equivalent to a thouſand arguments!——On this ſubject I would

would wiſh to ſay a few words on the debauchery that uſually attends county elections, eſpecially ſuch as are conteſted.

Much has been ſaid of late years on the ſubject of inſtructing Members of Parliament. No condition would be more juſtifiable than to demand of all the candidates a promiſe that they would not, by encouraging dabauchery, ruin the health, deſtroy the induſtry, and corrupt the morals, of thoſe people for whoſe intereſt they profeſs ſuch an anxious concern, and to whoſe ſervice they are ſo profoundly devoted. I believe it will not be thought going too far to affirm, that very few indeed have it in their power to repair, by any political conduct of their own, the miſchief done by a conteſted election. No combination among the electors could be more truly patriotic, than one which tended to refuſe ſupport to every candidate that attempted to promote his intereſt by ſuch means.

In the former part of this eſſay, I have mentioned ſome of the ill effects that follow violent exertions of labour or exerciſe, which I truſt are ſufficient to ſhew the imprudence of ſuch ſtrained efforts. I ſhall only add here, that ſuch trials ſhould not be encouraged by premiums or other means, either by private perſons or public ſocieties. It would be far preferable to encourage conſtant and perſevering

induſtry

induſtry and good execution of work, than exceſſive labour and fatiguing exertions of ſtrength.——I ſhall conclude this chapter with ſome pieces of general advice.

Thoſe who employ ſervants in agriculture ſhould encourage them to be careful of their health, and to make it a point of conſideration. A proper prudence in this reſpect is perfectly conſiſtent with induſtry, and is indeed the moſt neceſſary circumſtance towards the execution of a great quantity of work.—It is common with agricultural ſocieties to give premiums for the greateſt number of children; but this ſhould always be conjoined with another condition, that the children ſhould be healthy, and this laſt circumſtance ſhould preponderate againſt the other.

The ſituation of *farm-houſes* is a matter of great conſequence as it regards health. The greateſt care ſhould be taken to place them in dry ſituations with a deſcent from them every way, and upon a gravelly ſpot, or at leaſt ſuch a one as is free from ſprings that riſe to, or near to, the ſurface of the earth. Care alſo ſhould be taken to place the repoſitories for dung and other manure at ſome diſtance from the houſe, and this caution ſhould alſo be extended to the hog-ſtyes and poultry-yards. The neceſſary attention to the feeding theſe animals does not allow

the

the diſtance to be conſiderable, but ſtill does not require their being ſo near as we generally ſee them placed. It is needleſs to expatiate on ſuch a ſubject, or to attempt to prove that air impregnated with ſuch filthy exhalations muſt be injurious to health.

Even good plans for farm-houſes of different ſizes, according to the number of inhabitants, would contribute in no ſmall degree to general welfare. The bed-chambers in farm-houſes are in general too low and confined, and the whole building too ſmall; this occaſions too many people to be crouded together, a circumſtance always very unfavourable to health, and the moſt common ſource of contagious diſorders. Good water is alſo a circumſtance of great moment. If this can be had from any ſpring that riſes to the ſurface of the earth, it is commonly preferable to ſuch as is drawn from a conſiderable depth; but ſuch a choice is not always in our power. If pond-water be uſed through neceſſity, it ſhould be previouſly put into ciſterns or reſervoirs, covered at the top, and there ſuffered to ſettle. Care ſhould be had that the water be taken from a large pond, with a ſtony or gravelly bottom, and not ſubject to become putrid. Thoſe who drink water of this kind ſhould beware of ſwallowing the eggs or ſpawn of animals, leeches particularly, which ſometimes have produced, it is ſaid, diſagreeable ſymptoms.

Cleanlineſs

Cleanlineſs of the perſon is of greater importance to health than is generally imagined, and ought to be particularly encouraged among the lower ranks of people, eſpecially thoſe employed in this way. Nothing ſeems more likely to contribute to this ſalutary purpoſe than a due obſervation of Sunday; this precept is not only conducive to religion, morals, and civilization, but alſo to health.——It is well obſerved by Mr. Addison, that Sunday clears away the ruſt of the whole week:—an expreſſion which may be underſtood to extend to cleanlineſs as well as other conſiderations, and indeed appears to have been ſo intended by the amiable author in the paſſage referred to.

On the Cure *of the* Diseases *incident to an* Agricultural Life.

I now come to the laſt part of this eſſay, which is to ſpeak—*Of the Cure of the Diſeaſes to which Agricultural Perſons are ſubject from their way of Life:* and here I muſt remind the reader, that the preſent treatiſe is not meant to be a diſcuſſion of the ſubject at large in a medical way, but only to contain ſome plain hints and directions of the practical kind, which I believe to be juſtified by reaſon and experience.

Perſons

Perſons employed in daily labour of a healthy kind, and living on coarſe food, naturally become robuſt and athletic, of a firm fibre and denſe blood. Hence inflammatory complaints are in ſuch habits more common than thoſe of the putrid kind; and ſuch as are attended with low ſpirits and other hypochondriacal ſymptoms, are rarely met with. Evacuations may of courſe be uſed with more ſafety among ſuch people than among the effeminate inhabitants of populous towns.

Bleeding in the fevers that occur among country people, is for the moſt part neceſſary, eſpecially in ſuch as are attended with local inflammation, as pleuriſy, peripneumony, or inflammation of any of the viſcera. In ſuch caſes, twelve, fourteen, ſixteen, or even twenty ounces of blood, may, and often ought, to be drawn at one time. The quantity, however, cannot be determined by any general rule, but muſt be regulated by the age, ſtrength, ſex, and conſtitution of the patient, but principally by the urgency of the ſymptoms. If the internal pain be very acute, the ſkin hot and dry, and the pulſe exceed 110 beats in a minute, a large bleeding is generally neceſſary, eſpecially if any other ſymptom of a fatiguing or dangerous kind, as a violent cough, or ſhortneſs of breath, be preſent.

It is proper here to remark, that as ſoon as the nature of the complaint is ſo far aſcertained as to prove bleeding to be indicated, it is of conſequence that ſuch operation be performed as ſoon as poſſible, and that a ſufficient quantity be drawn at one time. One plentiful bleeding will ſometimes ſubdue a diſeaſe at its firſt appearance, when if half the quantity only had been taken, it would have required perhaps to be repeated ſeveral times.——It often, however, and indeed generally happens, in fevers attended with local inflammation, that one bleeding, however judiciouſly managed in reſpect of quantity, is not ſufficient. In ſuch caſes we muſt be governed nearly altogether by the urgency of the ſymptoms, and when theſe indicate a farther evacuation to be neceſſary, we muſt proceed, not indeed without regard to other circumſtances, but nevertheleſs as conſidering them as ſubſervient only to the principal object. An attention to this circumſtance is eſpecially proper, when the parts that are the ſeat of the complaint are immediately neceſſary to life, as in inflammations of the brain, lungs, bowels, or any of the viſcera; in ſuch caſes there is no time to be loſt, and what many would think bold practice, is indeed the only means of eſcape.——It is proper indeed to be careful, that the complaint originally be of ſuch a nature as to require bleeding at all; and in this, it muſt be confeſſed, even the moſt acute perſons of the profeſſion have been deceived.——The intermittent fever ſometimes comes on with ſuch violent ſymptoms as to reſemble very ſtrongly an inflammatory fever. But a little time generally reſolves the difficulty, and the ſucceſſive and clear marked ſtages of *cold*, *heat*, and *ſweat*, are for the moſt part ſufficient to determine the nature of the diſorder, even before any intermiſſion takes place, and any neceſſity of beginning to treat it as a fever of a different kind. Even if it ſhould be miſtaken, and ſome blood drawn, this evacuation has been often found ſerviceable in the beginning of intermittents, when the ſymptoms are violent, and is recommended on ſuch occaſions by the moſt judicious practitioners. A careful examination of circumſtances will, for the moſt part, enable us to diſtinguiſh this diſorder at its firſt appearance.—Moiſt weather, and a ſeaſon of the year about either the vernal or autumnal equinoxes, the latter eſpecially, and the frequency of the diſorder in the neighbourhood, afford ſtrong preſumptions in favour of a fever being of the intermittent kind;—to which we may add, ſuch obſervations as may be drawn from the nature of the ſoil and ſituation, and the buſineſs or work in which thoſe attacked with the complaint had been employed, previous to its firſt coming on.

It is an opinion generally received, that if bleeding be omitted at the beginning of fevers, it is improper in their advanced ſtate, and this is in ſome meaſure true. Fevers that commenced with inflammatory ſymptoms often become putrid as they proceed, and bleeding is certainly improper in ſuch circumſtances. But I would obſerve, that this caution holds more ſtrongly with regard to the enfeebled inhabitants of towns, than for robuſt country men. I apprehend that bleeding, though certainly more likely to be of ſervice if tried at the beginning of the diſeaſe, is nevertheleſs proper at every period when inflammatory ſymptoms are preſent. This holds more ſtrongly in caſes of fever attended with local inflammation, as in pleuriſy, inflammations of the viſcera, &c. in which the propriety of bleeding at every ſtage, provided the ſymptoms are urgent, is univerſally acknowledged. It ſometimes happens in robuſt people, that the common inflammatory fever preſerves its original appearance nearly as long as life continues, and its change of type is not to be regarded ſo much as an indication that points out the propriety of a different method of treatment, as a ſign that all our attempts are likely to be in vain.

Topical bleeding is often of great ſervice in many diſorders as well as general bleeding, eſpecially in the removal of ſome troubleſome and diſtreſſing ſymptoms.

ſymptoms. The head-ache is frequently an attendant on fevers, and often continues when the heat, quickneſs of pulſe, thirſt, and other ſymptoms, are much abated; and may be often thus relieved.—One of the eaſieſt and ſafeſt methods of partial or topical bleeding is by the application of leeches. If 3, 4, 5, or 6 of theſe, be applied to the temples, in the caſes mentioned above, they will often procure almoſt immediate eaſe, and are perfectly ſafe in their application, as the quantity of blood each of them draws is very ſmall. Leeches may often be applied with great ſucceſs in many inflammatory complaints that ſhew themſelves externally, as rheumatic ſwellings, particularly thoſe of the face and cheeks, inflammations of the eyes, inner parts of the ear, &c. In every inſtance they ſhould be applied as near as poſſible to the part affected. Bleeding, however, though a powerful remedy, requires ſome judgment and caution in the application of it.——The fevers that appear among country people, though often inflammatory, are not always ſo. Putrid fevers, though ſcarcely natural (if ſuch an expreſſion may be admitted) to a country life, are nevertheleſs capable of being communicated by contagion, and in ſuch caſes bleeding is generally hurtful. In ſome inſtances the putrid and inflammatory ſymptoms are ſo combined, as to make it doubtful to which clafs of fymptoms we ought principally to attend, and in fuch cafes fome experienced perfon fhould be confulted; but in general the fudden debility of body, and dejection of mind, that ufually come on at the accefs of the fever, the red watery eye, and the tendency to perfpiration or other evacuations, fufficiently diftinguifh this complaint from thofe of the inflammatory kind. The nervous fever feems to be only an inferior degree of the putrid or malignant. It is principally diftinguifhed by the weaknefs and dejection of mind that attend it.

The ulcerated fore-throat is another complaint that may be found in every fituation, as it is capable of being propagated by contagion. The difference of this from the inflammatory fore-throat is now well underftood and generally known; but there is another complaint that refembles it very much, which requires a very different mode of treatment, of which I fhall fpeak hereafter. In both, however, bleeding is improper.——No cafe requires the ufe of the lancet more than the common cold, if attended with cough and pain of the breaft or fide. Thefe fymptoms, if neglected, frequently terminate in confumptions, which might eafily have been prevented by fome evacuation of this kind, joined to common care, whilft the complaint was recent.

An

An abſurd cuſtom prevails among the common people, of letting blood about the ſpring and fall of the year, whether they have any complaint that requires ſuch evacuation or no; this practice, however, ſo far from tending to prevent diſorders, contributes greatly to produce them. It cauſes an habitual plethora, impoveriſhes the blood and juices, and when done at the latter end of the year, is apt to diſpoſe the body to intermittent fevers, and, if often repeated, to dropſical complaints. Many other bad effects of this abſurd practice might be enumerated, but they are, I think, unneceſſary to mention.

Purging is a mode of evacuation, whoſe conſequences in inflammatory complaints are often important, though leſs ſo than bleeding, and I think leſs hazardous, if miſapplied. Almoſt every inflammatory complaint requires ſome operation of this kind; it is however in general proper to be preceded by bleeding, which is thought to make it more ſafe and effectual: but this rule admits of many exceptions, and is not neceſſary to be adhered to, except the tendency to inflammation prevail pretty ſtrongly.

It was formerly thought, that purgative medicines differ conſiderably in the nature of the diſcharge they produced; ſome being calculated to diſcharge water, others bile, others phlegm, &c.

 but

but modern practice does not admit of much difference in this respect. All purgatives evacuate the bowels, and, if powerful and stimulating, produce a watery discharge by the absorption they occasion from the lymphatic system. Notwithstanding this similarity in the effects of purgative remedies, they differ considerably with respect to the circumstances that attend their operation. Some purgatives are observed to stimulate the body and accelerate the pulse during their operation more than others, and this is an important circumstance to direct our choice of them, according to the purposes for which they are intended. Those that operate with least irritation to the system, especially to the circulation, are preferable in acute complaints; and nothing in such cases is better than a simple solution of the bitter purging salt in water. It is seldom rejected by the stomach, however unpleasant it may be to the taste, and its operation is effectual and takes place quickly;—a circumstance of great importance in such cases. From one to two ounces may be safely taken, dissolved in a pint of warm water, in all inflammatory complaints where purging is proper. If it should be necessary to repeat it in the advanced stages, when the feverish heat begins to subside, it may be taken dissolved in the same quantity of infusion of flowers of chamomile, which will

will conduce to support the tone of the stomach without obstructing the evacuation.——The same remedy is proper in such fevers as are attended with local inflammation. If given early in such complaints, it will generally procure a passage, being quick and effectual in its operation, and found by experience to be less liable to be thrown up than things much more pleasing to the taste. The use of this medicine is not only adviseable in continued fevers, but also in the beginning of intermittents, when the patient is strong enough to bear purging. But of this I shall say more when I come to speak of the use of the Peruvian bark in that disorder.

In the advanced state of all fevers, when the inflammatory disposition begins to abate, and a tendency to putrefaction to prevail, the saline purgatives in general are less proper than such as are of a more warm and stimulating nature. Rhubarb in such cases is more proper, as being warm and aromatic, at the same time that it is purgative. If it be required to be made warmer, one half or one third part of nutmeg, or any other spice, may be added; from 20 grains to 60 may be given for a dose, but this must vary according to circumstances.

It is a perplexing circumstance attending the giving purgative medicines internally, that we cannot

determine

determine the degree of their operation by the proportion or quantity that is taken. It generally happens that one half or two thirds of the ufual dofe will have little or no effect; whereas had the full quantity been taken, it would have produced a larger difcharge than might be defired. In cafes, therefore, where fome evacuation of the bowels is neceffary, and at the fame time we might be apprehenfive of any unpleafant effects from a large difcharge, it is proper to employ clyfters, which have an additional advantage, that their effect takes place in a much fhorter time than could be produced by any purgative medicine internally taken. When clyfters are adminiftered with this intent, there is no great neceffity to be very particular in their compofition. A pint of warm gruel or broth, with two fpoonfuls of fallad oil, or melted butter, a table-fpoonful of common falt, and the fame quantity of brown fugar, forms as efficacious and proper a clyfter as the moft laboured compofition.

It is proper, while upon this fubject, to caution againft the practice of giving purgative medicines internally, efpecially fuch as are of the heating or ftimulating kind, commonly called warm purgatives, to people who complain of pain in their ftomach or bowels; particularly if this pain be attended with heat,

heat, thirſt, or other ſymptoms of fever. It is much the ſafer practice to inject a clyſter of the kind beforementioned, and to repeat it if neceſſary, and to uſe external fomentations, than to enter precipitately on the uſe of purgative medicines, which, if they do not take effect, often aggravate the miſchief, by producing or increaſing a diſpoſition to vomit, and ſometimes totally inverting the periſtaltic motion of the inteſtines. If ſufficient ſtools can be procured by clyſters, the danger is generally over; but if that means of relief do not ſucceed, it is ſafer to apply to ſome of the profeſſion, who may beſt determine what method may be purſued.

It is of the utmoſt conſequence to mention, that when any complaint of violent pain in the ſtomach or bowels is made, eſpecially if ſuch pain be not accompanied by ſtools, we ſhould enquire firſt about the place in which it is chiefly felt, if that can be pointed out; and next, if it came on rather ſuddenly, or ſoon after performing ſome laborious work, eſpecially the lifting any great weight, or indeed any conſiderable exertion of ſtrength. If this be found to be the caſe, we ſhould carefully examine the belly, eſpecially that part neareſt to the ſeat of the pain; and if any ſwelling, or lump, however ſmall, be found, even of the ſize of a hazel-nut, we may be almoſt certain, that the cauſe of the complaint originates from thence, and that if it be *immediately* attended to, it may *probably* be relieved, at leaſt the preſent danger obviated; but that if it be neglected, the patient will almoſt infallibly die. The only remedy on ſuch occaſion is, to reſtore the portion of the inteſtine, which is thus protruded and compreſſed between the muſcles of the abdomen, again into the cavity of the belly; and if this be done *ſoon* after the accident, it produces no farther injury. But this muſt be underſtood of ſuch caſes only as have a quick attention paid to them, ſince if any delay is made, the danger increaſes very rapidly: even a few hours may determine the interval between ſafety and death.——If the patient be in the vigour of age and ſtrength, the conſequences of neglect are more to be apprehended, than if he were advanced in years, as the probability of inflammation and ſtricture upon the inteſtine is greater.

The apparent facility and celerity with which this operation is often performed, and its great ſimplicity, may induce ſome people to attempt the performance of it, who have had no inſtruction or experience relative thereto; but it is proper to caution againſt ſuch attempts, as much nicety of touch, and addreſs of management, are often requiſite; and if the part be rudely or injudiciouſly handled, the hazard of the diſorder is much increaſed. The

parts

parts where ſuch an accident is moſt liable to happen, are the navel and the groin, but this rule is by no means univerſal.

It is farther neceſſary to remark, that women, who in country buſineſs are often employed in lifting conſiderable weights, as of pails of milk, buckets of water, and ſuch like, are more liable to ruptures than men; and on that account it is highly neceſſary that whoever attends women labouring under any acute pains of the abdomen, ſhould make a ſtrict enquiry into the circumſtances under which ſuch pains originated, and particularly if there be any tumour in the groin, belly, or pudenda; and if there be ſuch, to be informed of what nature it is, before he goes any farther, or loſes any more of that time, which in ſuch caſes is ſo very precious.

To return now from this rather long, but I hope not uſeleſs, digreſſion.

Emetics are another claſs of medicines of the evacuatory kind, that are often of the greateſt importance, and whoſe uſe requires the particular attention of thoſe who give advice to ſick perſons. It is not meant here to give a general account of the cauſes or circumſtances in which emetics may be adminiſtered with propriety, but only to give a few hints

hints relative to the cafes that are moft likely to occur in fuch fituations, in which thofe remedies may be ufeful.

Firft, then, emetics are indicated in cafes where from imprudence or negligence any thing has been fwallowed, that we have reafon to believe would be fpecifically injurious by its continuance in the ftomach. Thus if any poifonous plant, root, or berry, as of henbane, dropwort, nightfhade, or fuch like, has been incautioufly or ignorantly taken, our principal fecurity depends on fuch poifonous fubftance being evacuated as foon as poffible, and this can only be done with fafety by means of emetics. A fcruple or half a drachm of powder of ipecacuanha, together with a grain of emetic tartar, may be fafely given on fuch occafions to an adult perfon, and worked off with a ftrong infufion of chamomile flowers, or of root of horfe-radifh. This accident is moft likely to happen to children, with whom the fame remedy may be tried; fome diminution in the dofe may be proper; but in fuch dangerous cafes it is better to give a full dofe, and the rather as by the quicker and more effectual operation of a larger quantity, the emetic fubftance itfelf is fooner and more completely difcharged, and in general with greater eafe to the patient, than if a fmall dofe had been employed. The fame remedy may be

taken

taken when the ſtomach is overloaded by exceſs of food, or by any victuals that diſagree.

With the ſame intention emetics may be given when by ſome violent debauch the ſtomach is deluged with ſtrong liquors, and the inſenſibility may be ſo great, that it may be apprehended life is endangered. A quick and effectual evacuation of the ſtomach is ſometimes of great conſequence in ſuch caſes, and ſerves to reſcue ſuch unwary perſons from impending deſtruction. When the ſtupor prevails ſo far as to prevent ſwallowing, a few grains of emetic tartar conveyed to the back part of the tongue will find its way into the ſtomach, and in moſt caſes, where that organ retains any ſenſibility, produce vomiting. A few grains (three or four for inſtance) of blue vitriol may be uſed for the ſame purpoſe, and in deſperate caſes is preferable, as poſſeſſing a ſtronger emetic quality.

It muſt however be obſerved, that it is not always either adviſeable or ſafe to give vomits to remove ſubſtances that have got into the ſtomach, that we apprehend may do miſchief by means of their mechanical ſtimulus, as pieces of bone, pins, or other ſharp or pointed bodies, that may have been ſwallowed. The contraction of the ſtomach that neceſſarily attends vomiting may, if the ſubſtances

be

be not difcharged, aggravate their bad effects, and caufe mifchief by preffing on fuch fubftances; which might not have happened, had they been left undifturbed.

But the ufe of emetics is not confined merely to cafes where we defire to empty the ftomach, on account of any foulnefs fuppofed to be lodged there. They are often of the greateft fervice when given at the coming on of feverifh complaints, whether thefe be intermittent or continual fevers. In both of thefe, it is ufual for fome degree of naufea, or ficknefs of the ftomach, to accompany the cold fit, which it is proper to encourage, fhould it not come to an actual vomiting. An infufion of chamomile is often fufficient for this purpofe; but if that fail to excite a complete difcharge, a fcruple of ipecacuanha in powder will affift the difcharge fufficiently, effectually, and fafely.

Emetics are often of fervice in the common catarrhous cold, when the glands of the throat and fauces are deluged with mucous phlegm, which is often very difficult to be fpit up. In thefe cafes, an emetic often acts in the moft powerful manner in unloading the glands, and promoting general expectoration.

About

About ten years ago, a diſeaſe appeared in the Midland counties, much reſembling the ulcerated ſore-throat, but differing from it in reality, and requiring very different remedies. This was called the ſore-throat, attended with ſcarlet fever, and raged principally in the ſummer and autumn, in hot and dry weather, and attacked principally robuſt and vigorous people. Vomiting in this diſorder proved a very effectual remedy, and required to be frequently repeated, during the heat of the diſeaſe, even, in bad caſes, as far as twice in twenty-four hours. Should the ſame complaint again become epidemic, the early adminiſtration of emetics will probably be of the utmoſt conſequence, and ought to be carefully attended to. It was found neceſſary to uſe ſuch as were of a powerful kind, otherwiſe little benefit was received.

The above are far from being the only caſes wherein emetics are uſeful, but ſuch only as occur to me wherein they may be adminiſtered without hazard of being injurious, and have a probability of being of ſervice. It will be proper here to offer ſome *Cautions relative to the Doſe and Management of Emetics.*

Firſt, then, I apprehend, that it is a miſtaken notion that gentle emetics, as they are called, are

milder

milder in their operation than the more powerful. A ſmall quantity of ipecacuanha often cauſes a moſt troubleſome nauſea and retching, for a long time together, owing to its not poſſeſſing a ſtimulus ſufficiently ſtrong to cauſe a compleat evacuation of the ſtomach. A ſtrong emetic, on the other hand, by clearing the ſtomach in a few efforts, is itſelf diſcharged, and of courſe gives no farther trouble. A ſcruple of ipecacuanha in general, operates with much leſs pain and fatigue than five or ten grains, and the operation is ſooner over. It is proper to add a portion of ſome antimonial preparation to the ipecacuanha: a grain or two of emetic tartar, or a drachm or two drachms of antimonial wine, ſerve the purpoſe equally well. They are of ſervice in clearing the ſtomach more completely than ipecacuanha would do if given alone; and on the other hand, the ipecacuanha cauſes the antimonial medicines to operate with greater certainty as emetics, which would otherwiſe often go off by ſtool.

Another caution I would recommend is, to wait patiently for the operation of the emetic, and not to attempt by any mechanical means, as tickling the throat with a feather, or with the finger, to cauſe retching before the ſickneſs is ſufficiently ſtrong to excite vomiting freely. It is better even to repreſs the

the firſt motions to vomit, and wait till they become ſufficiently ſtrong to be effectual.

It is ſomewhat remarkable, that the addition of antimony to ipecacuanha, though it certainly cauſes a more full evacuation of the ſtomach, ſeems to retard the operation in ſome meaſure. Ipecacuanha given alone generally cauſes ſickneſs in the ſpace of fifteen or twenty minutes; whereas, if antimony be added, that effect ſeldom takes place under half an hour or forty minutes, and often longer.

Another caution is, for thoſe who take emetics not to load their ſtomach with large quantities of warm inſipid liquor, under the notion of working off the vomit; warm chamomile or horſe-radiſh tea, or a mixture of both, is preferable to gruel or warm water, as not relaxing the ſtomach ſo much as thoſe weak taſteleſs liquors.

Laſtly, it ſhould be conſidered by all, that the habit of taking emetics is of itſelf very weakening to the ſtomach and powers of digeſtion: occaſionally uſed, they are in moſt caſes a ſafe and powerful remedy, but frequent repetition cauſes them to be leſs beneficial, and in time hurtful.

Medicines that cauſe ſweat, called in medicine *Diaphoretics*, are next to be conſidered. The uſe of these, though not so general as was formerly thought, still forms an important indication. The common catarrhous cold is more effectually relieved by promoting this evacuation than by any other means, and the same was observed of the influenza, when that complaint was epidemic some years ago. Rheumatic complaints are also benefited by it, and many other slight febrile complaints. Nothing, however, has been more mistaken, than the most effectual means of producing this discharge. External heat is generally thought necessary; but it has been discovered of late years, that the body may easily be made too hot to sweat, and that there is often no method more powerful than by diminishing the heat of the body when too great, and that it is often necessary, in order to promote perspiration, to take off part of the bed-clothes, diminish the heat of the room by removing the fire and opening the windows, and to give cool liquors to the patient.—— Nothing succeeds better with this view in the common catarrhous cold, which requires some discharge by perspiration as much as any complaint whatever, than to bathe the feet at night for a quarter of an hour in water made about blood-warm, and to take a moderate dose of spirits of hartshorn in a pint of warm whey, gruel, or infusion of some garden herbs, as balm, mint, &c. This is perhaps the ſafeſt method of any; as whatever the nature of the diſeaſe may turn out afterwards, no injury can ariſe from what has been done.

In rheumatic caſes, it may be neceſſary to employ diaphoretics of a more powerful kind, and for this purpoſe *Dover's Powder* is frequently given, and often with good effect. It is given from five to fifteen grains, and may be continued every night or every other night for ten days or a fortnight, if the diſcharge by the ſkin be not too great, and the painful ſymptoms continue.

It is proper here to ſpeak a few words on the ſubject of that popular remedy, *Dr. James's Fever Powder*. This is well known to be an antimonial compoſition, leſs ſtimulating to the ſtomach and bowels than emetic tartar, and on that account preferable where any permanent effect is deſired. It often acts as an emetic or a purgative, as well as a diaphoretic; but the laſt effect is, I think, more common. In fevers of the inflammatory kind, and ſuch as are commonly found in country places, it is, if given with any tolerable caution, an excellent remedy, taking off the feveriſh ſpaſm, unloading the ſtomach and bowels, and as it were giving an opportunity for the exertions of nature. It is beſt given at the beginning of feveriſh complaints, be-

 fore

ſore they alter their tendency from an inflammatory to one that is putrid. If the inflammatory ſymptoms are violent, it is ſafer to uſe ſome previous evacuations by bleeding, and a clyſter; after which from five to twenty grains may be given, according to the ſtrength of the patient and urgency of the ſymptoms. It is given with moſt advantage in caſes wherein the patient is able to bear conſiderable diſcharge by ſtool; but it is remarkable that theſe ſeemingly diſtreſſing operations are ſoon recovered, and the patient appears in many caſes the ſtronger on their account. The ſickneſs alſo cauſed by this medicine, however uneaſy to bear whilſt it laſts, generally leaves the ſtomach in a ſtate fit for the receiving of nouriſhment; an inclination for which is, in many inſtances, one of the firſt marks of benefit received from the uſe of this remedy.

It ſometimes happens, that this medicine, though given to its full quantity, produces no ſenſible effect of any kind. Whether this be owing to any defect in the preparation, or to any inſenſibility in the nerves of the ſtomach at ſuch times, I cannot determine. It is however in ſuch circumſtances neceſſary to forward its effects, ſince if it remain inactive in reſpect of producing ſome evacuation or

other,

other, it ſeldom is of any ſervice. If the ſtomach appears to be loaded, a ſcruple of ipecacuanha may be given, joined to a grain of emetic tartar, which generally takes effect, and ſometimes ſeems to excite the action of the powder. If we wiſh to determine its action downward, an ounce or ſix drachms of the bitter purging ſalt may be taken, and a clyſter of broth and common ſalt thrown up.

As to the mode of exhibiting this medicine, I have before obſerved, that it may be taken from five to twenty grains at a doſe, and is moſt conveniently given in ſomething of a viſcid conſiſtence, as pulp of roaſted apple, currant-jelly, or the like. If put into any thin fluid, as tea, it is apt to ſink to the bottom, being of conſiderable ſpecifick gravity, and indiſſoluble in any watery fluid. It has been a great injury to medicine, that this preparation ſhould have been kept ſo long as an empyrical ſecret. It has been by that means extolled in complaints in which it had no ſalutary efficacy, and was even liable to prove injurious. It is impoſſible that any medicine can be ſuited to every kind of fever. Some require immediate and large evacuations; in others, ſuch a treatment tends infallibly to deſtroy the patient; and there is no doubt that the indiſcriminate recommendation of this remedy, which generally acts as a powerful evacuant, muſt have been the cauſe of many lives being ſacrificed to pecuniary intereſt. Such a remedy, however ſuited to diſorders where a quick and powerful evacuation is required, is utterly improper in caſes where the powers of life are much reduced, and where the utmoſt attention to ſupport the ſtrength of the patient is neceſſary. This is always the indication in fevers of the low, nervous, and contagious kind, and is frequently the caſe in the advanced ſtate of fevers in general, whatever might be their tendency at their firſt appearance. Even in the rheumatiſm, which is of an inflammatory nature, though often chronical in point of its duration, this medicine, though often ſerviceable at the beginning, muſt not be continued very long, as it is found, like the other antimonial preparations, to injure by long uſe the tone of the ſtomach and powers of digeſtion.

The high, and as it might juſtly be called extravagant price of this remedy, which bore ſcarcely any aſſignable proportion to its intrinſic value, has cauſed its uſe to be leſs general among the poor than humanity would deſire. That objection is now, however, ſuperſeded; the powder being now ſold at Apothecaries-hall, for leſs than one twentieth part of its former price; and this powder is found, on the moſt impartial examination, to be fully equal in every reſpect to that ſold under the denomination of *James's Powder*.

Diuretic Medicines form a claſs of remedies, whoſe effects would be very deſirable, were they not ſo precarious. No diuretics that we know are much to be depended upon for certainty of operation, eſpecially in ſuch caſes as we moſt deſire to have it. Thoſe that are ſafeſt, and leaſt offenſive to the ſtomach, are the ſweet ſpirit of nitre, and the ſweet ſpirit of vitriol, which may be taken in the doſe of a tea ſpoonful in a glaſs of water, or other cold liquor, once or twice a day, and continued for two or three weeks. Some of the vegetable infuſions, as of horſe-radiſh and muſtard-ſeed, will ſometimes produce the ſame effect, and may be continued, if ſucceſsful, a long time without injury to the conſtitution or health in other reſpects. The infuſion may be made by pouring hot water on the muſtard-ſeed bruiſed, and horſe-radiſh freſh ſcraped, and letting them ſtand together a few minutes. An ounce of each of the ingredients is enough for a quart of water; and about a quarter of a pint of this infuſion may be taken twice a day.——The above infuſion, or one ſimilar to it, is often uſed with ſucceſs in the ſwelling of the belly and legs, which often ſucceeds obſtinate intermittent fevers, and is generally attended with thick turbid urine, which is ſecreted in ſmall quantity. This preparation, though apparently ſimple, is as likely to ſucceed as many others that are more compounded, and may be ſafely tried in all caſes where the urinary ſecretion is defective.——It is proper, however, to caution againſt the giving diuretic medicines of any kind, when any pain, or heat of urine, accompanies the diminution of its quantity. In ſuch caſes, opiate and emollient remedies are proper, joined with ſuch as abate inflammation.

Having thus ſpoken of medicines that produce evacuation, I ſhall now ſpeak of thoſe which ſtimulate, and call the powers of life into action.

Stimulant Medicines may be conſidered in practice as of two kinds; one of which tends to give a permanent ſupport to the vital powers, the other tends to excite their action in a more temporary manner. Of the former of theſe, wine, when good, is perhaps the moſt generally uſeful in caſes of emergency. It is now found that, in low and putrid fevers, wine may be given with great advantage in larger quantity than was formerly thought practicable with ſafety, even to two or three bottles in 24 hours. Nay, larger quantities have been adminiſtered, but it has been found that even a proper remedy may be over-doſed, and that ſuch quantities as I have above ſpecified, ſhould be cautiouſly ventured on, and not without attentively obſerving the effects of each doſe that is taken. The beſt way of giving it is, I think, in ſmall quantities and frequently, and as freſh from the cellar as poſſible, perfectly cool, and without any admixture.

In fevers, where the ſkin is moiſt, with a ſcalding heat to the touch, the pulſe quick and low, the eyes moiſt or watery, the ſtools looſe and fœtid, thirſt great, tongue foul, reſpiration difficult, and ſpirits depreſſed, there the uſe of wine is adviſeable, and is indeed the principal remedy on which we muſt depend. The indication for wine is ſtronger, if any ſpots of a blue or purple caſt appear on the body, or if a low muttering delirium come on, attended with faintneſs. Life then depends on active and quick exertions. Moſt fevers that are contagious are of this kind, particularly that called the *Jail Fever*; and the ſame method of treatment is neceſſary in ſome meaſure in moſt acute fevers that laſt beyond eight or ten days, without ſome evident ſigns of abatement. The marks whereby we may judge wine when adminiſtered to be of ſervice are, a ceſſation or diminution of the pain in the head, or delirium, diminution of the heat and clammy ſweat, and by the patient being in better ſpirits, and entertaining hopes of his recovery. It often happens, that fevers

of

of this kind, when they begin to abate, aſſume ſomewhat of an inflammatory appearance, the ſkin becoming hot and dry, and the pulſe full and quick. Theſe ſymptoms are not unfavourable, and generally abate of their own accord. They indicate, however, that wine ſhould be more ſparingly given, if not totally laid aſide, during their continuance.

If wine cannot be had, or not in perfection, or is not reliſhed by the palate, good malt-liquor may be ſubſtituted in its room; and I have ſeen porter tried with the beſt effect in a caſe of this kind. The patient drank about three quarts a day for ſeveral days, and it ſeemed to agree better than wine or any other medicine, and was, after the ſecond day, the only remedy adminiſtered. I have ſome reaſon to think, ale, or ſtrong beer, might be uſed with ſimilar effect; but have never had any perſonal experience of their efficacy. The Peruvian bark is uſed with the ſame intention in the ſame diſorder, and with good effect. But it is now thought that wine is full as powerful, and much more eaſily adminiſtered, as being more grateful to the palate:—a thing of great importance where the frequent repetition of a medicine is neceſſary. The uſe of the bark is, therefore, in a good meaſure ſuperſeded in putrid fevers, except where the throat is ulcerated; in which

which complaint it has been found by experience to be particularly useful.

The principal use of the *Peruvian Bark* is in the intermittent fever, the returns of which it is well known to be very efficacious in preventing. It is best given in substance, and most conveniently in form of an electuary made up with any syrup, and with the addition of some spice, as a little nutmeg, or cinnamon, in powder, to each dose. If the patient be strong, and the body costive, a small quantity (a drachm for instance) of Glauber's salts, or the bitter purging salt, may be added to the three or four first doses of the bark, which generally opens the body and promotes urine; but if the disorder be advanced, or the patient weak or in years, such addition is less proper. If the bark purges, such tendency must be moderated, which a few drops (two or three for instance) of liquid laudanum in each dose generally does very effectually; and when that disposition is conquered, the laudanum must be omitted. The bark must be given in considerable quantity when employed to cure an intermittent. It is to little purpose to give to a grown-up person less than an ounce in twenty-four hours, and often double that quantity. It may be given in doses of two scruples or a drachm each, or about the bulk of a large nutmeg of the electuary every two hours

on

on the day of intermiſſion, and repeated every day for ſeveral days, if the fever does not return. After the intermiſſion of three or four periods of the paroxyſms, we may diminiſh the quantity, and give it only every four hours, taking care to give a doſe a little before the time of day that the return of the complaint may be moſt probably expected. If the complaint does not return, the quantity may be in the ſpace of a week or ten days ſtill farther diminiſhed, but it muſt not be left off entirely for the ſpace of at leaſt ſix weeks. If the diſorder has had ſeveral returns, if it be an autumnal ſeaſon, and the weather rainy, if the fits return every day, or with an interval of two days, or if the patient be weak and emaciated, more caution and attention to the regular adminiſtration of the bark will be neceſſary.

It would be a deſireable circumſtance, if that kind of the Peruvian Bark, called the *Red Bark*, were to be had genuine; but at preſent there is reaſon to think, that it can ſcarcely be procured. In an obſtinate caſe of an intermittent, that fell lately under my care, I had an opportunity of trying the effects of the calamus aromaticus, which given in combination with the Peruvian bark, in the proportion of one part to two of the bark, ſtopt the progreſs of an intermittent that had reſiſted the bark

taken

taken alone. It is proper during a courſe of the bark to uſe a moderately liberal diet; but all exceſs, either in meat or drinks, is carefully to be avoided.

It is an old prejudice that ſubſiſts even to the preſent time, and among ſome of the medical profeſſion, that intermittent fevers ſhould not be too ſoon ſtopt, but ſuffered to go on through ſeveral paroxyſms, before the bark ſhould be given. It was thought that ſeveral diſorders, particularly indurations of the liver, jaundice, meſenteric obſtructions, and even rheumatic complaints, were produced in conſequence of the bark being, as it was thought, prematurely given. But it now appears, that theſe complaints were the conſequences of the diſorder, being ſuffered to continue too long, not of its being too ſoon ſtopt, and that the beſt method of preventing them is to interrupt the courſe of the fits as early as poſſible by a ſteady and reſolute uſe of the proper remedy. This caution ought to be carefully attended to, and enforced by thoſe who give advice to people in country places, as the prejudices in favour of the fits being ſuffered to continue, are often very ſtrong.

Bitter Medicines, ſuch as the flowers of chamomile, roots of gentian, and centaury, are, in a good meaſure, ſimilar in their effects to the Peruvian bark.

bark. They are, however, lefs effectual in the cure of intermittents and diforders of a putrid tendency, but better fuited to a weak ftate of the ftomach and organs of digeftion. A ftrong infufion of any of the above-mentioned articles, with a little rind of the Seville orange, makes a bitter preparation as efficacious as any, and as pleafant as fuch a medicine can well be expected to be. A quarter of a pint of this taken twice a day for a week, fortnight, or three weeks, will often be of great fervice in diforders of the ftomach unattended with fever. Infufions of this kind are alfo convenient vehicles for the adminiftration of fome other medicines. I have before mentioned, that faline purgatives may be given to advantage diffolved in an infufion of flowers of chamomile, and the fame is true of fuch diuretic medicines as are of fmall bulk. The dulcified fpirits of vitriol, and of nitre, may be given in this way, as may falt of tartar when ufed as a diuretic.

There is likewife another clafs of ftimulant remedies, which feem to act more generally on the fyftem, though they fometimes excite particular fecretions. The infufion of *muftard feed* and *horfe-radifh*, before recommended as a diuretic, is of this kind, and is often given with advantage in cafes where the fecretions in general are languid and flow.

It

It may be taken with confiderable advantage in rheumatic cafes of long duration, where the pain is rather tedious and troublefome than acute, and attended with little or no fever. This medicine may be continued a confiderable time, feveral months for inftance, with lefs injury to the health and conftitution in general, than might be expected from the long ufe of fubftances, whofe fenfible qualities are fo powerful.

Stimulant applications of the *external* kind are next to be confidered. The principal of thefe are *Bliſters*. The proper ufe of thefe in many inftances is attended with much nicety, and of courfe not a fubject of my prefent treatife, which is only to give cautions, and to recommend the ufe of fuch remedies as may be applied with fafety in fuch circumftances as are obvious to common underftandings, independant of medical knowledge.—Neverthelefs, there are, I think, many opportunities of their being employed with fafety and probability of advantage, in circumftances that require no great medical knowledge to difcover. Thus the ufe of blifters is adviſeable in all internal pains, whether of the breaft, fide, or belly, attended with fever. In fuch cafes, after bleeding, a blifter, applied as near to the feat of the pain as poffible, is a fafe, and in general, if put on early after the commencement of the difeafe,

an

an efficacious remedy; which may, if neceſſary, be repeated with perfect ſafety.——In caſes, likewiſe, where cough and pain of the breaſt, though unattended with much fever, are ſymptoms, I have ſeen the beſt effect from ſmall bliſters repeatedly applied to the part where the pain was felt; and believe, if they were oftener tried when theſe ſymptoms are but recent, might prevent many complaints of the lungs, which a ſhort neglect renders fatal. I have found it the moſt eaſy, as well as effectual method of uſing this remedy, to apply it at going to bed, and, if it has riſen, to remove it in the morning, and ſuffer it to heal up, and if neceſſary to repeat it. This is leſs troubleſome, and I think more effectual, than a perpetual bliſter.

Bliſters are uſeful in pains of the head accompanying fever, or where any tendency to vertigo or delirium appears. If applied at the firſt appearance of theſe ſymptoms, which are always alarming, they are often of the greateſt ſervice, and ſafe in their application. They are moſt ſerviceable in ſuch caſes, if applied to the head when freſh ſhaved; but as that cannot always be done, eſpecially with women, they muſt be put on between the ſhoulders. The ſame remedy is often uſed in ſome local inflammations, partly of the external kind. Thus in the inflammation of the eye, or inner part of the

ear,

ear, bliſters behind the ears frequently bring, after other evacuations have been uſed, great relief; which is likewiſe the caſe in violent pains in the cheek and face.

The ſtrangury ſometimes follows the application of a bliſter. This however, though troubleſome, is ſeldom of any ſerious conſideration, as it is moſtly relieved by drinking plentifully of any mild warm diluting liquor, as milk and water, infuſion of linſeed, ſolution of gum arabic in an infuſion of the root of marſh-mallows, and ſuch like. It is thought to conduce to the prevention of the ſtrangury, in thoſe ſubject to it, to cover the bliſtering plaiſter, as far as the flies extend, with a piece of gauze or muſlin, and to ſpread the margin with the gum plaiſter, to ſecure its adheſion. By this management, the whole of the flies are taken off when the bliſter is dreſſed, which contributes to prevent the abſorption of their acrimonious particles, which are thought to be taken into the circulation by being ſuffered to remain on the raw part. This precaution is of ſervice, but not always quite effectual, and ſhould not be uſed when a quick operation is required, as it abates the activity of the cantharides. It is cuſtomary with ſome to ſuffer the bliſtering plaiſter to remain on the part twenty-four hours, but I think the time ſhould be meaſured by the effect; and if a blister be raised in a third part of the time, as is often the case if the plaister be good, there is no occasion to trouble the patient with it for a longer time, which often gives unnecessary pain, and is much more likely to cause strangury, than if it were removed earlier.

The dressing of blisters is worthy attention.—The best of any is a simple plaister of white wax and olive oil, melted together by a very gentle heat, and spread thin on a rather fine linen cloth. This keeps the part from the air, and does not stick to it, or cause any irritation. It is proper, previous to the application of a blister, to examine the surface of the plaister, that it may be sufficiently moist; since, if it be too dry, it will often produce no effect whatsoever. If it seem dry and unpliable, it should be gently warmed before the fire, and moistened, first with a little spirits of wine or brandy, and then with a little olive oil or fresh butter. We must be cautious in practice of applying blisters at all in cases of the putrid kind attended with fever, and where inflammations of the urinary passages are present.

Blisters, however, are not the only forms in which external stimulants may be employed with advantage. It sometimes happens, that it may be convenient to employ a remedy of quicker operation, as in violent pains of the head, delirium of fevers, apoplectic or paralytic feizures, and the like. In fuch diftreffing circumftances, it has been frequently found that ftimulant applications to the lower extremities have been of great fervice, and proved a fafe as well as an effectual remedy. Muftard feed bruifed, or in fine powder, as the flour of muftard, is the beft application. If this be mixed with an equal quantity of ftale bread grated down, and made into a rather moift pafte with vinegar, it will form a cataplafm of a proper confiftence for the purpofe here intended. If this be fpread about a quarter of an inch thick on a piece of leather or linen cloth, and applied to the foles of the feet, or in extremities to the whole of the feet, it almoft immediately produces a violent burning fenfation, and fometimes, though not always, an inflammation of the part; and now and then it raifes a blifter. The time it fhould be fuffered to remain upon the part, muft be meafured principally by the effects it produces. It fhould not be removed immediately on the firft abatement of the fymptoms, nor need it be kept on till they have entirely ceafed, as the fenfation continues a confiderable time after the cataplafm is removed.

In lefs arduous cafes, as in fixed rheumatic pains of the hip, fhoulder, or other parts, a plaifter of brine has been applied with advantage, as being leſs painful than muſtard, and leſs apt to bliſter than cantharides. In the tooth-ache, and pains of the face, a convenient temporary application may be made by mixing a little black pepper ground into powder, with as much brandy or other ſpirits as will make it into a ſoft paſte, which is to be ſpread on leather, and applied to the face. This produces a conſiderable ſenſation of heat, but without any great uneaſineſs, and ſeldom bliſters, tho' it is often very efficacious in removing the pain of the part.

The foregoing application is very uſeful at the firſt coming on of a ſore-throat, if laid upon the outſide under the chin, and moiſtened again with ſpirits as it becomes dry. I never knew it bliſter, though it is frequently of great ſervice. A mixture of ſpirits of hartſhorn, with olive oil put upon flannel, and laid to the throat, is often adviſed; but I think it not ſo effectual as the former.

Medicines that eaſe pain, and procure reſt, are the next to be conſidered; theſe are of ſeveral kinds, but Opium, by its greater efficacy, and more convenient exhibition, has ſuperſeded in a great meaſure all the others.——Did opiates produce no other effects than thoſe above aſcribed to them, it would be unneceſſary to give any directions relative to their

their uſe, farther than to determine the proper doſe; but the operation of this remedy is not ſo ſimple, but requires attention to regulate, and, in ſome inſtances, to counteract ſome of its effects. Opium may be ſafely and properly adminiſtered in moſt caſes of violent pain, attended with none, or but little fever or inflammation. Thus it is the principal, and indeed almoſt the only remedy to be depended on, in thoſe dreadful fits of pain which often attend the paſſage of a ſtone or gravel thro' the urinary paſſages. In ſuch circumſtances, opiates may be given with conſiderable freedom, in proportion to the exceſs of pain which is neceſſary to be alleviated, not only for the purpoſe of procuring eaſe to the patient, but alſo to allow the ſtone to paſs, which ſeldom happens unleſs the pain and conſequent ſpaſm can be abated. Twenty, forty, or ſixty drops, or any intermediate quantity of the tincture of opium, or of liquid laudanum, may be taken in twenty-four hours, according to the urgency of the ſymptoms. Larger doſes have been given, but they are not without hazard, ſince as the pain is from the nature of the complaint liable, and indeed often does ceaſe ſuddenly, from the paſſage of the irritating ſubſtance, the opium then is left to exert its full effect, unchecked by the ſtimulus of the pain, and aided by the diſpoſition to

 ſleep,

ſleep, which naturally comes on after the ceſſation of great torment. This is ſaid, in ſome inſtances, to have produced fatal effects, the ſleep proving mortal. On this account it will be prudent, after giving as large a doſe of an opiate as can ſafely be done, to endeavour to allay the pain by other means, as fomentations, warm baths, &c. until the effect of the opiate be gone off a little, and a ſecond doſe may be given with ſafety. If the complaint be attended with vomiting, as thoſe of the nephritic kind frequently are, a larger doſe may be ventured on, if we find that what has been before given has been thrown up; but we muſt not conclude, that the effect of opiates is *quite* loſt, even though they ſhould be rejected from the ſtomach. Their ſtay is generally ſufficient for them to ſhew ſome ſigns of their ſpecific qualities. In caſes where opiates are proper, and where there is any great nauſea or tendency to vomit, it is more convenient to exhibit this medicine in a ſolid form; and it is found by experience, that the ſmaller the bulk of the remedy, the greater is the probability of its being retained upon the ſtomach. A ſmall pill, therefore, made of a grain of unſtrained opium, without any other admixture, may be uſed in place of thirty drops of tincture of opium, to which it is fully equivalent; and this may, if neceſſary, be repeated once in

twenty-

twenty-four hours. If the vomiting be ſo violent as to ſuffer nothing, however ſmall, to be retained upon the ſtomach, opium may be conveniently administered in a clyſter. Forty drops, or a moderate tea-ſpoonful, which is generally regarded as equal to a drachm in meaſure, may be mixed with about half a pint or leſs, of broth, gruel, or warm milk, and injected as a clyſter, and retained, if poſſible, ſeveral hours.

It is always proper, that the body be kept, if poſſible, in a rather lax ſtate during the uſe of opiates. If, therefore, any coſtiveneſs be preſent, it is adviſeable to inject a clyſter of a moderately opening kind, previous to the giving of the opiate, which makes the operation of the latter ſafe, and leſs liable to affect the nervous ſyſtem. If the uſe of opiates be neceſſary to be continued, it is proper to adminiſter occaſionally ſome internal medicines of a mildly purgative kind, as opiates generally render the body coſtive. The precautions juſt mentioned are equally applicable to ſuch bilious diſorders as are attended with great and often exquiſite pain about the pit of the ſtomach, without fever, and generally without any increaſe of pulſe, and are produced by the gall-ſtones ſticking in the ducts which convey the bile from the liver and the gall-bladder into the inteſtines. It ſhould, however, be conſi-

dered,

dered, that opium, in both the above-mentioned cafes, is only a temporary relief; and though it often affifts the paffage of the obftructing body, yet is of no fervice to prevent the return of the diforder, and therefore fhould be taken only when great pain, and other urgent and diftreffing fymptoms, render its ufe neceffary.

The ufe of opium is in no inftance more ftrongly manifefted, than in the violent purging and vomiting that often comes on towards the latter part of the fummer, or during the autumn, and is called the Cholera Morbus. It may not be proper to give opiates immediately on the accefs of the diforder, but after we may reafonably fuppofe the ftomach and bowels to be cleared of their proper and natural contents, and little but bile, water, or mucus, paffes, it is time to adminifter opiates, efpecially if the retching to vomit, diftention of the ftomach, and griping pains, be violent. In fuch cafes there is no time to be loft, and opiates are often the only refource. They may be given either in a liquid or folid form. The liquid opiate takes effect fooner, but is more liable to be thrown up, on which account we fhould endeavour to make it as acceptable to the ftomach as poffible. About a fpoonful of warm fimple mint-water, or of peppermint, is as likely to make it ftay on the ftomach as any thing

I know,

I know, and the ſmaller the quantity of fluid ſwallowed with it, provided it be ſufficient to diſguiſe the taſte, the more proper.

Opium is likewiſe proper in the ſimple diarrhœa or purging, that often comes on towards the cloſe of ſummer. This, though ſometimes ſalutary when moderate, often continues ſo long as to exhauſt the ſtrength and weaken the tone of the ſtomach and bowels. In ſuch caſes it is often neceſſary to combine the opiate with ſome cordial aſtringent, among which I think cinnamon the beſt. If an ounce of cinnamon in powder be made into an electuary with any ſyrup, and the bulk of a ſmall nutmeg taken three times a day with four, five, or ſix drops of tincture of opium added to each doſe, it forms a powerful and ſafe remedy in autumnal fluxes.

Opiates, judiciouſly adminiſtered, might often prevent many of the bad conſequences that follow violent colics, the iliac paſſion, and inflammation of the bowels. If a ſufficient doſe of tincture of opium, 20 or 30 drops for example, or, what might perhaps be more proper, a grain of unſtrained opium in a pill, were to be given as ſoon as the pain becomes violent, and before any vomiting has come on, it might allay the pain, and make way for the operation of clyſters; and would be preferable, in

my

my opinion, to the exhibition of ſtrong purgatives taken by the mouth, which, if they fail of producing an evacuation downwards, as they often do, cauſe vomiting, and aggravate all the other bad ſymptoms. If, however, the pain be violent, and accompanied with heat or thirſt, it will be neceſſary to let blood, which is perfectly compatible with the operation of opium. This medicine is not only uſeful on account of its own ſpecific qualities, but alſo as a corrector of thoſe of other medicines.

It is not uncommon for the Peruvian bark to act as a purgative, which, in ſome of the moſt arduous circumſtances in which it is given, as in intermittents, and fevers of a putrid tendency, is apt to diſappoint its good effects. In ſuch caſes a few drops of tincture of opium, added to each doſe, generally prevents the bark proving purgative, and of courſe ſuffers it to continue in the body long enough to be ſerviceable. This need not in general be continued long, as after a few doſes the bark will uſually loſe its purgative quality, and may be taken alone.

I ſhall conclude this head with a caution relative to the uſe of ſuch remedies; which is, that the taking of them is very apt to inſinuate itſelf, and to become habitual, eſpecially in thoſe who are occaſionally ſubject to painful diſorders. If often uſed,

they

they become almoſt neceſſary, as ſleep cannot be procured without them. Thoſe, therefore, who are obliged to take opiates occaſionally, ſhould make it a point of conſequence not to uſe them except when they are obviouſly neceſſary, and to leave them off as ſoon as that neceſſity no longer exiſts. It may be troubleſome at firſt, but ſleep will return in time ſpontaneouſly, if the party have but reſolution to perſevere. The long continuance of opiates requires an increaſe of the doſe, which produces coſtiveneſs, indigeſtion, general weakneſs, and a tribe of nervous ſymptoms, very ſimilar to thoſe which are the conſequences of dram-drinking, which the taking of opiates in large quantities very much reſembles.

I have thus finiſhed what I have to ſay on the uſe of the medicines, which are calculated to anſwer the principal indications of cure. The reader will perceive eaſily that this extends only to diſeaſes of a certain deſcription, and reſpects in them only thoſe remedies, the propriety of whoſe uſe is obvious to any perſon of ſenſe and tolerable education, without aſſiſtance from profeſſional ſtudy or information. I hope I have advanced nothing that is liable to miſlead, and I truſt that what is adviſed, will tend to make what future medical operations may be neceſſary more ſucceſsful. Phyſicians often juſtly

lament,

lament, and often when it is too late, the imprudent measures that have been taken previous to their being employed. To obviate this in some measure, is the principal intention of the present essay, which I have put into the present form, as being comprised in smaller compass than if I had treated of the diseases separately, and as I think more easy to be understood.

Some general directions relative to the *treatment of Sick Persons*, which could not so properly be introduced under the several indications, are here added.

Cleanliness is a matter of the greatest consequence to the cure both of acute and chronical disorders. Every person who is indisposed ought to wash the face and hands, and feet occasionally, with at least equal regularity as in health. The ease and comfort this affords to sick persons, those especially who labour under acute disorders, can scarcely be imagined, except by those who have experienced it.—Change of linen is a highly necessary article. Every person ill of a fever ought to have clean linen for the head and body every day, and clean sheets every three days, or oftener, if the perspiration be large. Many causeless

causeless fears formerly prevailed, and still subsist, concerning the airing of linen. It is necessary undoubtedly that this should be dry, but this is best insured by its being exposed when perfectly clean to a fire, and dried by that means only. Linen that has been worn, or sheets that have been lain in, with a view to airing them, are unfit for sick persons, as they are in reality fouled and damped by such absurd care. The room itself should be swept daily, and every offensive thing removed as soon as possible, and nothing suffered to remain in the room that is not immediately necessary to the patient. Whatever food or drink he does not consume should be removed immediately after as much as is necessary be taken, and no person suffered to take food in the room except the sick person.——It is necessary that the utmost care be taken that the victuals, and whatever besides be prepared for the sick, be dressed with the greatest regard to cleanliness. The stomach in such cases is always delicate, and it is of the greatest consequence to indulge it in this respect.

Change of Air and *Coolness* are nearly connected with Cleanliness, and equally necessary to be attended to. Every person confined to their bed with any feverish complaint, should have the door and window of the room opened for a quarter of an

hour

hour twice at leaſt in twenty-four hours. If the weather be very ſevere, the curtains may be drawn ſo as to prevent the current of air from blowing on the ſick perſon; but if the weather be mild or warm, the door and windows ſhould be open thro' the day and even the night. There cannot be the leaſt neceſſity why the air ſhould be warmer for a perſon ill of a fever than for a perſon in health, but many reaſons why we ſhould wiſh it cooler. To keep a ſick perſon's chamber well aired, (I mean here not by fires, but by opening the doors and windows) contributes not only to the benefit of the ſick perſon, but to the ſafety of the attendants.—— Many fevers, which were at firſt ſimply inflammatory, have become by heat, confinement of air, and other improper treatment, putrid and contagious. It is obvious that if the complaint originally be of a putrid tendency, theſe cautions become doubly neceſſary.

The proportion of bed-clothes is a circumſtance of great moment. Theſe muſt be meaſured partly by the age of the patient and nature of the complaint, but principally by the feelings of the ſick perſon. I have often obſerved, that much of the reſtleſſneſs attending fevers, which is ſo troubleſome and fatiguing a ſymptom, and ſo exhauſting of the ſtrength of the patient, is owing to the quantity of bed-clothes. Too much heat naturally produces a desire to change the heated place for one that is cooler; but if the body were not uneasily hot, no such impatience would take place.

Quiet is another important article. Officious curiosity is apt to make many persons intrude upon sick people, who have very little real concern on their account. This should by all means be discouraged, and no more persons admitted to the chamber of the sick than are necessary to attend him. The admission of others tends only to foul the air, increase the heat, and prevent the rest of th sick.

The proper administration of food is a matter of the utmost concern. At the beginning of inflammatory complaints, provided the patient be young and robust, some abstinence may be proper; but in the advanced state of all disorders both acute and chronical, it is necessary to pay great attention to the article of food. As long as any considerable degree of fever remains, it is proper to use a vegetable diet. Milk boiled with bread, bread and rice puddings, roasted apples, and the like, are all proper; and for drink, toast and water, whey, or (if it be good) small beer; but no stronger liquor, and least of all distilled spirits. A most absurd and unhappy

notion

notion ſtill prevails among many of the lower people, that ſick perſons are in want of ſomething cordial to ſupport their ſtrength, and keep up their ſpirits; in conſequence of which they often, in ſpite of any directions to the contrary from thoſe who are better qualified to judge, give them ſome ſtrong fermented liquor, as ale or ſtrong beer, and ſometimes rum, brandy, gin, and the like. It is a melancholy reflection, that numbers fall victims to this odious opinion, which, as well as many other inſtances of impertinent interference with the ſick, ought to be diſcouraged as much as poſſible.

The mode of giving food to a ſick perſon is worthy regard. It ſhould always be in ſmall quantity, and no more ſhould be brought into the patient's ſight than it may be expected he will be able to take at a time. This ſhould be done often in the day, and even during the night, and without waiting for its being aſked for by the ſick perſon, who is often able to take food when he has not ſpirits to aſk for it. It muſt not however be preſſed with any importunity, which is more likely to excite diſguſt than appetite.

The ſupport of the ſpirits of a perſon labouring under diſeaſe, is as neceſſary towards his cure as the

the adminiſtration of medicines. Every perſon that is ill, ſhould be comforted with hopes of recovery, and cheerful proſpects of life. To foretell a perſon's death in his preſence, who is then ill of an acute complaint, has no ſmall influence in verifying the prediction. Even thoſe whoſe profeſſion leads them to recommend religion to others, ſhould be careful of dwelling too much upon gloomy ſubjects, and giving people diſpiriting ideas of their ſituation.—— Repentance and amendment of life are no doubt in many inſtances neceſſary to be adviſed, but great care muſt be taken to adminiſter, together with advice, that greateſt of all cordials—*Hope*.

I have before mentioned, that an opinion prevailed among the lower ranks of people, that bleeding at certain times of the year was a ſalutary practice, whether any immediate neceſſity appeared to make it proper or not. The ſame abſurd notion prevails with reſpect to the taking purgative medicines. It is needleſs to ſay more here, than that ſuch habits are extremely improper to be commenced, and ſhould be by all means if poſſible prevented. If, however, they have been begun, we muſt not precipitately direct them to be omitted, but to diminiſh the quantity of blood taken, and to omit the purgative, and in time lay them both aſide altogether. I am inclined to think, that the almanacks, in which ſuch advice has been for many ages prepoſterouſly inſerted, have been the principal cauſes of ſuch abſurd notions being carried into practice for ſo long a courſe of years. I ſee it has been of late omitted in ſome, and hope the others will follow the example.

A prejudice ſubſiſts among many people of the lower ranks, againſt every remedy that does not operate upon them in ſome ſenſible manner as an evacuant. They do not meaſure its good effects by the change it produces upon the health, but by its increaſing their natural diſcharges. This is an unfortunate prepoſſeſſion, as ſeveral of the moſt effectual remedies act for the moſt part without any ſenſible alteration in the animal ſyſtem, ſave the ceſſation of the diſorder. This is the caſe in general with the Peruvian bark, when given as a cure for the intermittent fever, in which, if medicines of the evacuatory kind were to be joined with the bark, they would, unleſs very gentle in their operation, fruſtrate the good effects of the principal remedy. It is proper on this account, whenever medicines of this kind are given, to forewarn thoſe to whom they are adminiſtered, that they are not to expect from them any other effect than an abatement

ment of the diforder which they were intended to remedy: a condition furely fufficient to fatisfy any reafonable perfon.

The common people are too apt to eftimate the efficacy of medicines, as they do that of other things, by their pecuniary value and their fcarcity. They have no idea that Providence has made the moft ufeful things in medicine, as well as food, cheap and common, and that expence in fuch articles is oftener neceffary to flatter and comply with effeminate delicacy, than to add to the real efficacy of a remedy. The poor who are in hofpitals do not receive, in proportion to their numbers, lefs relief than the rich in their fplendid apartments; though in the former cafe nothing be conceded to prejudice, fancy, and caprice; and in the latter, it makes the moft important confideration. It is incumbent, therefore, on all who take the charge of the lower people when fick, to combat this miftaken opinion, and to endeavour, if poffible, to convince them, that the beft remedies are in many inftances the cheapeft.

Thofe who take the charge of fick perfons fhould be cautious that the fame courfe of medicines be not continued too long a time together. It fhould be underftood, that medicines (at leaft the greateft part of them) are more calculated to *restore* health than to *preserve* it. We should therefore be careful to recommend, to persons in health, to be contented with the happiness they enjoy in that respect, and not to attempt to improve what cannot be amended, but may easily be impaired. Some ignorant people are prepossessed with a notion that it is *wholsome*, as it is termed, to drink several infusions of herbs, as of flowers of chamomile, of centaury, and several others. But such trials are not only unnecessary, but likely to be injurious. The taking of bitters in large quantities, for a long time together, hurts the tone of the stomach, instead of mending it, as was found by fatal experience of those who took the Portland Gout Powder, which destroyed nearly all who tried it. This powder was nothing else than such bitter herbs as are commonly drank in tea, or brewed with malt liquor in the form of purl. What is here said, is not meant to insinuate that bitters *properly* and *moderately* used, are not very useful remedies. It is the excess only that is censured.

Another reason why we should be upon our guard against continuing the use of the same medicines for a long time is, that it is apt to introduce that most destructive of all habits, *Dram-drinking*. Many of of the tinctures recommended in this way are little elſe than drams concealed under a medicinal diſguiſe, and as ſuch ſhould be with equal caution avoided, as far as reſpects their becoming habitual. I have more than once ſeen a habit of this odious kind introduced among women, particularly by theſe means. It is not ſo likely to happen to the lower ranks, as to thoſe who employ them, for whom this caution is principally meant.

The laſt piece of advice I ſhall offer reſpects QUACKEREY.

Perhaps there is nothing diſgraces the police of this country, more than the numerous impoſitions of this kind that are daily advertiſed. Scarcely any one of them has not only a greater certainty of ſucceſs aſcribed to it, but is alledged to be infallible in a greater variety of diſorders than are curable by all the articles of the Materia Medica taken collectively. Some of theſe boaſted remedies are merely frivolous and inert, but others are violent and dangerous in their operation, and highly improper to be truſted to ſuch perſons as thoſe who are thus raſhly encouraged to take them in an indiſcriminate manner. A ſolution of arſenic is ſaid to have been the baſis of a late ſpecific for fevers,

and

and I am well informed has in ſeveral inſtances deſtroyed the patient. Theſe inſtances however, are carefully concealed, whilſt every eſcape is carefully recited as a cure, owing to the remedy ſo given.

No piece of humanity would be greater than to preſerve the ignorant and uneducated of the lower ranks from ſacrificing their health and money to unfeeling fraud and intereſted knavery.

POSTSCRIPT.

REPORT RESPECTING THE TRIALS OF PLOUGHS IN MARCH 1788.

IN confequence of the premiums offered for afcertaining the cheapeft and beft plough, for the common practice of hufbandry in thefe parts of the kingdom, a field of ftrong old ley ground, part of Barracks Farm, near Bath, was felected as proper for the teft of experiments.——The perfons who declared themfelves candidates for this trial, were,

1. JOHN BILLINGSLEY, efq; of *Afhwick-Grove*, with a double coulter-plough, to be drawn by fix oxen, in yokes and bows.

2. Mr. HENRY VAGG, of *Chilcompton*, with the Norfolk plough, having two wheels, and one handle, to be drawn by two horfes abreaft, and guided by the ploughman without a driver.

3. Mr. JOHN THOMAS, of *Keynfham*, with a light fwing plough of his own improvement, to be drawn by four fmall Welch oxen, in yokes and bows.

4. Farmer SULLY, of *Midford*, with a fingle plough of this county, fomewhat lightened and improved, having a fmall wheel under the beam, in a line with the coulter, and to be drawn by three horfes lengthwife.

5. Mr. GEORGE FLOWER, of *Midford*, with a fingle plough, commonly ufed in this county, and to be drawn by three horfes lengthwife.

6. Lord

6. Lord WEYMOUTH, with the common ſingle Wiltſhire plough, to be drawn by three horſes, two abreaſt, and a ſingle leader.

For theſe candidates, ſix parallel pieces of ground were marked out, near one acre each, and all the ploughs were to begin at the ſame time, and to plough their reſpective lots at pleaſure; but as nearly as poſſible four inches deep, and eight inches wide. —On a previous trial of the ſoil, the Norfolk plough, from having only one handle, and the man not being uſed to plough ſtiff ley-land, was found unequal to the conteſt, and Mr. VAGG declined it. Mr. GEORGE FLOWER alſo on account of inferior workmanſhip, occaſioned by the ill-conſtruction of his plough, declined. Thus the conteſt began with only the other four. Before a judgment could be formed of the probable iſſue, Lord WEYMOUTH's plough was broken againſt a point of a rock juſt beneath the ſurface, and conſequently thrown out: —the trial then was confined to three.

At the end of three hours and four minutes Mr. BILLINGSLEY's plough had finiſhed its lot. At the end of five hours and five minutes Mr. SULLY's had finiſhed: and Mr. THOMAS's at the end of five hours and a half. The latter ploughed about half his lot with the four ſmall oxen, and the remainder with the addition of a ſingle horſe, the ſoil being found too ſtiff for the ſtrength of the oxen.——The committee of judges was compoſed of five practical farmers, three from Wilts, one out of Somerſet, and one from the county of Gloceſter.

On

On a full examination and comparison of the goodness of work, it was the opinion of the majority of the committee, that the double-coulter plough had the preference, for general purposes of husbandry, laying the furrow more flat than the others, and consequently exposing more new surface to the influence of the elements, and preventing more completely the growth of grass and weeds between the furrows. The want of a wheel to the swing-plough occasioned an unevenness of furrow and depth, which rendered the ploughing rather inferior, on that soil, to the work of the horse-plough; though it appeared much inferior to what it might probably have been, had the regularity of a wheel aided the excellent turning up of the mould-board; for which reason the owner was requested to pursue his improvement of a plough, which in several respects promised considerable utility as an ox-plough on level soils; more especially as it was asserted by Mr. THOMAS, that on such a soil, in summer fallow, his man had ploughed, with the same plough and oxen, an acre in three hours and forty minutes.

Finally, the premiums were awarded thus:—

1. The first premium of six guineas to JOHN BILLINGSLEY, esq; with a gratuity of one guinea to his servant.

2. The second premium of four guineas to Farmer SULLY, with a gratuity of half-a-guinea to his servant.

3. The third premium of two guineas to Mr. JOHN THOMAS, with a gratuity of a smock-frock to his servant.

And ſuch was the evident comparative ſuperiority of Mr. BILLINGSLEY's double coulter-plough, drawn by ſix oxen, that ſeveral gentlemen and farmers from different parts have in conſequence determined to work oxen inſtead of horſes; and have given orders for the making of double ploughs to the amount of ſix or ſeven in number. One gentleman, who was an umpire on the occaſion, and who occupies ſeveral large farms, having been accuſtomed to keep on one of them ſix horſes and two ploughs, being convinced that a double-coultered plough and ſix oxen would completely do the work of the farm, determined to make ſuch a regulation immediately. Thus the Society may have the ſatisfaction of hoping, that from a continuation of ſimilar public trials, improvements will be made in the leſſening of expence in huſbandry, from which, among numerous other cauſes, the propoſed good conſequences of their zealous endeavours will reſult.

To the foregoing Statement of Facts relative to this Trial of Ploughs, we ſubjoin the following Extract of a Letter written by a practical Farmer, who was preſent on the occaſion.

——Let us here pauſe, and take a comparative view of the expence of ploughing an acre of land drawn from the preceding trials.

The average price of keeping oxen, (including winter and ſummer food) I take to be 3s. per week; the

the calculation then in reſpect to Mr. BILLINGSLEY's plough will ſtand thus:

	£.	s.	d.
Six oxen, at 6d. each per day - -	0	3	0
Ploughman and driver - - - -	0	1	8
Wear and tear of plough, yokes, &c. -	0	0	4
Total	£.0	5	0

Let us ſuppoſe that $1\frac{1}{2}$ acre of ley, or $2\frac{1}{2}$ acres of ſtubble, or fallow land, be ploughed each day, the expence of the former will not then exceed 3s. 4d. per acre, and of the latter 2s. per acre.——Is not this improvement worthy the attention of all farmers?—And are not the thanks of the public due to the perſon who has been inſtrumental in bringing forward to view a reduction of expences in ploughing, which cannot be eſtimated at leſs than 2s. 6d. per acre?

Farmer SULLY's account may be thus ſtated:

	£.	s.	d.
Three horſes, at 9d. per day each - -	0	2	3
Ploughman and man driver* - - - - -	0	2	4
Wear and tear of plough, harneſs, ſhoeing, &c. ſay only - - - - - - - - -	0	0	5
	£.0	5	0

But as a boy might have guided the horſes, I will conſider the expence at 4s. 6d. per day; and herein

* This is properly remarked, as from an ill-judged policy in the farmer, a man-driver was employed, inſtead of a boy, that the horſes might be conſtantly kept to their greateſt ſpeed.

I think

I think no partiality is ſhewn to the ox-plough.— If one acre of ley, or $1\frac{1}{2}$ acre of ſtubble, or fallow, be ploughed in a day, the expence will then be 4s. 6d. the former, and 3s. per acre the latter.—Superiority of Mr. BILLINGSLEY's plan in both inſtances 1s. per acre.—And this ſuperiority would be ſtill greater in a compariſon with Mr. THOMAS's, were it not that the unſkilfulneſs of his ploughman, and the ſmallneſs of the oxen, rather preclude a ſtrict compariſon.

Now farmers are in general quick-ſighted enough in many caſes wherein their intereſt is concerned; ſurely, therefore, they cannot ſhut their eyes in this inſtance, but muſt adopt the uſe of a plough ſo vaſtly ſuperior to thoſe in common uſe.

END OF VOL. IV.

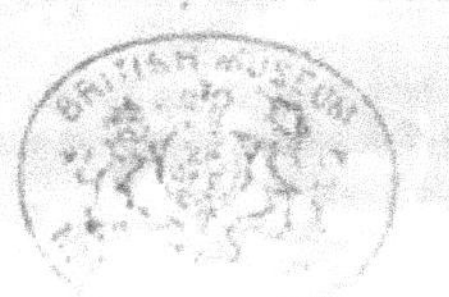

www.ingramcontent.com/pod-product-compliance
Lightning Source LLC
Chambersburg PA
CBHW020902210326
41598CB00018B/1751

* 9 7 8 3 3 3 7 1 2 5 2 2 6 *